Recent Results in Cancer Research

Fortschritte der Krebsforschung

Progrès dans les recherches sur le cancer

10

Edited by

V. G. Allfrey, New York · M. Allgöwer, Chur · K. H. Bauer, Heidelberg · I. Berenblum, Rehovoth · F. Bergel, Jersey, C. I. · J. Bernard, Paris · W. Bernhard, Villejuif N. N. Blokhin, Moskva · H. E. Bock, Tübingen · P. Bucalossi, Milano · A. V. Chaklin, Moskva · M. Chorazy, Gliwice · G. J. Cunningham, London · W. Dameshek, Boston M. Dargent, Lyon · G. Della Porta, Milano · P. Denoix, Villejuif · R. Dulbecco, San Diego · H. Eagle, New York · R. Eker, Oslo · P. Grabar, Paris · H. Hamperl, Bonn R. J. C. Harris, London · E. Hecker, Heidelberg · R. Herbeuval, Nancy · J. Higginson, Lyon · W. C. Hueper, Bethesda · H. Isliker, Lausanne · D. A. Karnofsky, New York · J. Kieler, København · G. Klein, Stockholm · H. Koprowski, Philadelphia · L. G. Koss, New York · G. Martz, Zürich · G. Mathé, Paris · O. Mühlbock, Amsterdam · W. Nakahara, Tokyo · G. T. Pack, New York · V. R. Potter, Madison · A. B. Sabin, Cincinnati · L. Sachs, Rehovoth · E. A. Saxén, Helsinki W. Szybalski, Madison · H. Tagnon, Bruxelles · R. M. Taylor, Toronto · A. Tissières, Genève · E. Uehlinger, Zürich · R. W. Wissler, Chicago · T. Yoshida, Tokyo

Editor in chief

P. Rentchnik, Genève

Springer-Verlag Berlin Heidelberg New York 1967

Radioactive Phosphorus
in the Diagnosis
of Gastrointestinal Cancer

Robert S. Nelson

With 61 Figures

Springer-Verlag Berlin Heidelberg New York 1967

*Robert S. Nelson, M.D., Professor of Medicine, General Faculty, University of Texas;
Internist, Chief, Gastroenterology Service, Department of Medicine, The University of
Texas M. D. Anderson Hospital and Tumor Institute, Houston, Texas/USA*

Sponsored by the Swiss League against Cancer

ISBN-13: 978-3-642-99928-4   e-ISBN-13: 978-3-642-99926-0
DOI: 10.1007/978-3-642-99926-0

Softcover reprint of the hardcover 1st edition 1967

Foreword

The diagnosis of cancer in the inaccessible regions of the gastrointestinal tract is difficult at best. Neoplasia frequently advances insidiously and largely without the patient's knowledge. Ideally, simple survey tests applied periodically to those segments of the population considered most susceptible should be available. For all practical purposes such means of diagnosis are nonexistent. Those who specialize in gastrointestinal cancer must, therefore, do the best they can. The best consists of many means, all good in themselves, but often subject to failure or misinterpretation of results. Any aid which will give even a small amount of positive information to tip the balance for or against the diagnosis of cancer in an obscure situation must be considered of value to the gastroenterologist.

The material presented in this volume represents our experience with such an aid over the past eight years. The use of radioactive phosphorus (P^{32}) and a miniature Geiger counter to record the differences in beta emission over tumors as compared to normal tissue now appears clinically useful in the diagnosis of gastrointestinal malignancy where the organ is available for such instrumentation. An attempt has been made to present the findings as objectively and as specifically as possible so as to provide maximum assistance to other gastrointestinal oncologists as well as others who have only a general interest in the subject. From the few references in the literature, it may be seen that the method has not yet been extensively tested by others. Hopefully, this presentation may stimulate wider use and better evaluation.

As in all such projects, the individual most involved must owe considerable to those who have rendered consistent and invaluable support. The author wishes to express a deep debt of gratitude to RAYMOND G. ROSE, M.D., who first suggested the study, CLIFTON D. HOWE, M.D., for his continued help and interest, WILLIAM C. DEWEY, Ph.D., for technical assistance and collaboration without which the project would have been impossible, MARTHA YOUNGERMAN for her faithful and skillful recording of results, and to LOUISE KLARQUIST for the many hours she has spent in typing and editing the manuscript.

Table of Contents

General Principles Involved in the Use of Radioactive Phosphorus (P³²) in Cancer Diagnosis and its Employment Outside the Gastrointestinal Tract

The use of radioactive phosphorus (P³²) for the diagnosis of malignant neoplasia was initially suggested by basic research on phospholipid turnover in normal and neoplastic tissues. The work of JONES et al. (1940) demonstrated that the phospholipid turnover of tumors bears a greater resemblance to the more active tissue, such as liver, kidney and intestine, as well as the fact that each type of tumor displays a characteristic type of activity, the rate of turnover differing among different groups. These workers also noted P³² was retained in greater concentration and for longer periods in tumors than in other tissues of the host, in this particular instance the laboratory mouse. The tumor uptake was high and rapid in the early intervals after administration of P³², and there was a pronounced capacity for retention of the isotope for long periods. It was felt that this characteristic served to distinguish tumor from normal tissue. MARSHAK (1940) similarly found that while the fraction of P³² bound by mice liver nuclei remained constant, the relative concentration in tumor nuclei of the same animal rose to more than twice that of the tumor tissue. At 48 hours, the specific activity of tumor nuclei was found to be more than four times that of liver nuclei. ERF and LAWRENCE (1941) after studying mice carrying various neoplasms and concluding that wherever there was leukemic or neoplastic infiltration, there was a higher uptake of P³² than in uninvolved tissue, injected radioactive phosphorus into humans dying of various malignant tumors and measured tissue content after death. In patients with seminoma and melanoma, the malignant tissue retained much more P³² than other tissues. In neuroblastoma, Ewing's sarcoma and Hodgkin's disease, P³² content was essentially the same as that of the more rapidly metabolizing organs such as liver and kidney. They concluded that the majority of the malignant tissues or those infiltrated with malignant cells retained as much or more radiophosphorus per gram wet weight than other rapidly metabolizing tissues of the same patients.

MARINELLI and GOLDSCHMIDT (1942) measured the concentration of P³² in normal and pathological skin in one patient with cutaneous melanoma and two with mycosis fungoides who were undergoing treatment with this agent. After administering therapeutic doses, they used a Geiger counter tube with cylindrical glass walls to record radiation over the lesions and normal skin, with the counter in direct contact to the tested area. In all three cases, P³² pickup and elimination rate, over the affected areas, were found to differ radically from those of the normal skin. Radioactivity over the melanoma nodules compared to the normal skin increased with the passage of time, but the same relative values were maintained in the two patients with mycosis fungoides. The amount of radiation in mycosis fungoides lesions appeared to correlate well with the degree of pathological malignancy. Low-Beer et al. (1946) first suggested using radioactive phosphorus in the diagnosis of malignant tumors in human beings. Citing the work of Marinelli and Goldschmidt as a basis for their study, they injected 300 to 500 microcuries of radioactive phosphorus intravenously, 24 to 48 hours prior to operation in 25 patients with breast tumors. Using a bell-jar type Geiger counter with thin glass windows having a diameter of 1.5 to 2.5 cm, they obtained surface measurements directly over the palpable breast tumor and over comparable areas of the opposite normal breast and other fleshy areas of the body at two, four, six, and 24 hours following injection. Surgical samples of the tumor and other tissues were dissected, weighed and ashed. Sixteen patients had carcinoma, and of these 15 showed a 25 per cent difference in count between the involved and adjacent areas of the same breast, and comparable areas of the opposite breast. The one false negative proved to be a mucoid carcinoma with relatively few cells. There were no false positive readings among the nonmalignant lesions. Quantitative determinations of the radioactivity of removed tissues showed a five to ten times greater uptake of radioactive phosphorus in cancer tissue than in any other tissue examined, except for the mucoid cancer, where contents of all tissue were the same.

These early clinical studies stimulated a gradually increasing interest in the use of radioactive phosphorus as a diagnostic, rather than a therapeutic, tool. The basic studies and later clinical work served to establish certain important principles in its use as well as the degree of accuracy which could be expected in different human organs. Insofar as the isotope is employed with Geiger counter tube scanning, it appears obvious that the pure Beta emissions cf radioactive phosphorus give it a distinct advantage over agents with Gamma radiation. The Beta rays are more specific, concentrate more in neoplastic tissue, and there is less background radiation from surrounding normal tissue. It was noted early that the better the counter tube covered the tumor to be tested, with little normal tissue at the periphery which might confuse the issue with additional radiation, the more striking was the differential obtained. The Beta rays have a maximum energy of 1.7 MeV and a maximum range in tissue of 8.0 mm, but the average energy is 0.7 MeV and the average range in tissue is 2.5 mm. While this penetration is quite adequate for accurate readings in most areas where it is possible to make close and fairly prolonged application of the Geiger tube, such as the skin and the eye, there is understandably considerable room for error where such approximation may not always be possible, as in the esophagus, stomach and rectum. In practical application it was also found that inflammatory areas might give positive readings, particularly when counts were obtained soon after giving the isotope intravenously. This effect, as might be expected from the basic studies, could be at least partly overcome by taking readings at a longer interval (20 to 48 hours) after giving the P³². The tumor counts obtained after these periods were not as high as those noted within one, two or four hours, but the differential between normal and neoplastic tissue was greater, and the false positive elevated readings from inflammatory lesions tended to subside to background, or normal tissue levels. With the basic principles established, and the probable worth of the method conceded, it remained only to devise proper instrumentation and procedure for the testing of different areas and organs of the human body.

Use of the P^{32} Test Outside the Gastrointestinal Tract

Ophthalmologic. From the first, the eye appeared to be a prime organ in which to assess the value of the P^{32} test. Despite the most careful examination, many eyes had been needlessly sacrificed because of a diagnosis of probable malignant tumor, the postoperative histologic diagnosis being that of hemorrhage, inflammation or simple retinal detachment. In addition, the eye appeared to be one of the most favorable sites for testing from the technical and anatomic viewpoint. It was accessible, essentially reproduced bilaterally for easy comparison of a diseased and nondiseased organ, and counts could be made accurately and for relatively prolonged intervals.

The test was initially developed by THOMAS et al. (1952), based on the previous use of P^{32} in the diagnosis and localization of breast tumors as described by LOW-BEER (1946). They injected 500 μc of P^{32} intravenously one-half hour before scanning with a small, thin-end-window Geiger counter. Counts were made at intervals of one-half hour, one hour, and one and one-half hours after injection of P^{32}. Tetracaine was used as a local anesthetic. Radioautographs were also made of resected tissue. Six of the eight cases tested were melanomas, and gave positive readings. The remaining two, in which no tumor was found, gave negative readings. Radioautographs confirmed the selective uptake over tumor. In subsequent clinical studies in 32 patients counts were taken up to a total of three days. It was noted that the differential ratio of tumor to normal counts reached a maximum very early (less than 15 minutes), and that the ratio then decreased for one to two hours following injection. After this, the ratio showed an increase which continued over the period studied. Anterior lesions gave the highest differentials, and posterior the lowest (THOMAS et al., 1953). Subsequent experimental work by this group employing animal tumors (1956, 1957) confirmed these results and demonstrated that inflammatory lesions, while giving an initial rise of radioactivity during the first 10 minutes, then subsided to background readings. In their most recent report (1965) they indicated that after study of 150 cases of ocular pathology, the ratio of uptake over tumor to normal in all accessible malignancies was greater than 1.6, whereas the ratio of nonmalignant lesions was less than 1.4. There was, however, no complete separation of the two groups. In 15 inaccessible tumors, ratios lower than 1.5 were obtained, but only one false positive was noted. The method was felt to approach 100 per cent accuracy in accessible lesions.

BETTMAN and FELLOWS (1954) carefully analyzed the factors involved in the P^{32} test in vivo and in vitro in quick-frozen human eyes. They noted that radioactivity was greatest over the most vascular tissues; that there was difficulty in obtaining counts over the posterior pole, and that certain avascular tumors, such as retinoblastoma, might actually produce lower counting rates than normal tissue. They concluded that the test might be valuable with carefully controlled conditions, a counter designed to be applied directly over the posterior pole, and with counts taken at longer intervals after injection of P^{32}.

EISENBERG et al. (1954) in two reports, using 500 μc of P^{32}, found in their first study that melanomas gave in general 30 per cent greater radiation than normal, whereas benign lesions gave less than this figure, and defined an increased uptake of 30 per cent as indicative of neoplastic change. They pointed out that positive uptakes might be given by inflammatory conditions, and recommended repeated mea-

1*

surements for both control sites and tumor area. In their second report after study of 123 cases, they changed their testing times to one, six, 24 and 48 hours, and modified their criteria for positivity to 1) one-hour uptake of 30 per cent or more in excess of normal; 2) a 24 hour uptake showing a percentage increase greater than the one-hour reading. It was found that a negative result indicated a nonmalignant lesion in 95 per cent of cases, and that a positive test was strongly indicative of malignancy.

In other tests, KENNEDY et al. (1954) reported that in 24 intraocular neoplasms, 11 nonneoplastic lesions showed no significant increase in radioactivity, while nine with various types of neoplasms gave diagnostic readings. Three cases of melanosarcoma showed no significant differential. They made no attempt to reach the posterior portion of the eye operatively, nor to apply the counter subconjunctively, and reported that a negative reading with a posterior lesion was not significant. SHAPIRO (1957) although using posterior counts with a curved counter applied subconjunctively to the sclera, drew attention to three false negatives in 24 cases, and concluded that the P^{32} test could be considered adjunctive only. DUNPHY et al. (1957) pointed out that the test could be improved in posterior lesions by incising the conjunctiva for direct counting. They recommended obtaining counts at 24 and 48 hours after injection of P^{32} for maximum accuracy, and stated that this method of diagnosis must be regarded as a diagnostic aid and not a diagnostic test. A high degree of accuracy was found by DONN and McTIGUE (1957) in 40 consecutive cases of suspected malignant melanoma. They found no false positives in 34 cases, and but one false negative. Five lesions however, were inaccessible to the probe counter. O'ROURKE et al. (1957) critically studied the method in the light of uptake studies on animal eyes, in which they noted wide difference in the paired eyes of an individual animal, as well as considerable (two-thirds) absorption of choroidal emissions by sclera, and higher concentrations in conjunctiva than in choroid. They intimated that it might be necessary to raise the upper normal figure of 30 per cent which had at this point been fairly well accepted. TUBWAY and PALIN (1958) briefly reviewed the subject of the P^{32} test, and noted that in general they had obtained positive results from primary malignant melanoma and secondary carcinoma in the eye, but had had no success in detecting retinoblastoma. RIORDAN et al. (1959) reported 10 correctly identified tumors of 13; the three false negatives were melanotic sarcomas, and were within range of the counter. In a good review and discussion of the P^{32} test in skin as well as eye tumors, LIDGETT et al. (1959) reported readings in 27 patients, 12 with cutaneous and 15 with ocular disease, and felt that the test was useful, a differential uptake of 1.3 being taken as significant. Their study also revealed a correlation between the P^{32} uptake ratio and the tumor nucleoprotein content based on measurements of volume and density of tumor cell nuclei. O'ROURKE and COLLINS (1960) studied the difficult problem of scanning tumors of the posterior choroid. Measurements were made under general anesthesia in seven patients with proven melanomata at the time of enucleation, 500 μc of P^{32} having been given 24 hours prior to surgery. By precise and complete scanning of the posterior eye, they found that with eight transconjunctival countings the malignancy was localized correctly in only one instance, and that the average focal increase of activity was only 25.3 per cent above the mean. With six transscleral countings, the tumor was localized correctly in all instances, and the average focal increase in radioactivity was 74 per cent. In a progress report on a large series of tests in ocular lesions, CARMICHAEL and LEOPOLD

(1960) noted an 86 per cent accuracy in 134 cases, 17 being incorrect. Their results with the posterior probe were not encouraging, but they felt that overall the test was a useful one since the degree of accuracy remained very stable over a period of years in a large group of patients. GOLDBERG et al. (1961) found an accuracy of 98 per cent in 103 cases (22 equivocal cases were excluded). They emphasized the necessity of using the transconjunctival approach in posterior choroidal lesions, and taking readings at 48 hours after injection of the isotope.

Within the limitations described by the workers in this field, it would seem that the P³² test is a valuable adjunct to testing the eye for the presence of malignant melanoma, and possibly metastatic cancer from other sites. Melanosarcoma and retinoblastoma apparently cannot be detected by this means. The posterior lesions are still difficult to test, and truly representative counts can be detected only by operative placement of the counter probe in most instances. The blending of background radiation with that from the tumor is confusing in small lesions, and the amount of inaccuracy furnished by overlying conjunctiva and sclera is still a point of argument. Nevertheless, the test is quite accurate when a definitely positive differential is obtained (30 per cent or greater), although a negative result does not eliminate the possibility of malignant tumor, particularly in the posterior eye. There is general agreement that the P³² test in ocular tumors should be considered as a diagnostic aid, and that all other methods of diagnosis should also be carefully weighed in reaching a conclusion as to enucleation.

Central nervous system. Radioactive phosphorus has been shown to play a limited but important role in the diagnosis of brain tumors. SELVERSTONE et al. (1949) first devised equipment and a method for the localization of tumors at surgery through a craniotomy opening or burr hole. In this original report, it was emphasized that approximate mapping of deep brain tumors had not presented a major problem when neurologic examination, electroencephalography and ventriculography were carefully employed, and that the study was limited to the more difficult area of precise location. Preliminary work had shown that when aliquots of tumor removed at intervals after injection of P³² were compared for radioactivity with white and grey matter removed at the same time, the radioactivity of tumor ranged from 5.8 to 110 times greater than that of adjacent normal white matter. The activity of gray matter was about one to two times that of white matter, but never approached that of tumor tissue. The Beta radiation of P³² penetrates only about 7.5 mm in brain or brain tumor, according to this study, which would allow location within this distance as a maximum. The first Geiger counters used were 5.0 mm in diameter, which were later reduced to 3.0 mm, since the former were considered too large and destructive for routine use. A preliminary group of 14 cases of brain tumor were selected for testing, in three of which the tumor was on the surface, and the remaining 11 within the brain substance. Radioactive phosphorus in the amount of 1.95 to 4.2 millicuries was given intravenously at varying intervals of 1.8 to 65.8 hours prior to operation (exact timing was apparently not felt to be important). After craniotomy, the counter was first inserted in a neurologically "silent" area to obtain control or background values, decontaminated by irrigation and moist sponging, and then successively introduced into suspected convolutions until tumor, shown by an increase in rate from 5.4 to 36.3 times that of the control area, was located. After encountering tumor, the Geiger tube was cleansed with hydrogen peroxide and alcohol in order to prevent possible implantation of malignant cells into normal areas. The results noted in this preliminary clinical trial were excellent; tumors were located in all 14 patients, although in one additional patient with a deep metastatic nodule, the probe never encountered the tumor. In a later report, SELVERSTONE and WHITE briefly listed their total experience with the method of a 2 mm diameter miniature Geiger probe in localizing brain tumors, employing P³² as a tracer in 114 cases, and K⁴² in 36 others. While excellent results were obtained with both agents, K⁴² differed from P³² in several essential respects. K⁴² provided useful differentials within a few minutes after injection, in comparison to P³², where a delay of up to 12 hours might be necessary. Since K⁴² had a half-life of only 12.44 hours, however, it was found necessary to obtain shipments at least twice weekly, whereas P³², with a half-life of 14.3 days, was much more convenient to use. Four cases were reported in detail in which it was possible to locate and delineate the position and extent of deep cerebral tumors by this method. Particular advantages noted were the ability to obtain a valid biopsy, to decide in advance whether the individual tumor was operable, and to scan for remaining tumor following resection of the gross lesion.

MORLEY and JEFFERSON (1952), noting the advances made in scintillation scanning for gross location of tumors with the skull intact, reported their results with the method of SELVERSTONE and co-workers in the delineation of brain tumors at craniotomy. Tests were made in 37 cases of over 200 tumors operated upon during the same period, and P³² scanning was reserved for instances when it was felt that information it might give was obtainable by no other means. Specifically, the test was found to be particularly valuable for the recognition of tumor tissue for biopsy, to map the extent and position of a subcortical tumor, and not so valuable in scanning the tumor bed following resection, since the counter apparently would not pick up small accumulations of cells left behind. In two negative tests, tumors were later found which proved to be fibrillary astrocytomas. Meningiomas, glioblastomas, and secondary carcinomas were found to emit Beta particles at comparable rates, while gross central necrosis in tumors showed reduced radiation, as did astrocytomas. Increased radioactivity was believed to be not due entirely to increased blood supply within a tumor. In general, the P³² test was rated as a valuable additional aid to the surgeon at operation in selected cases, particularly in the taking of a small biopsy specimen through a burr hole.

In subsequent reports, GARRITY and MATTHEWS (1954) found the counter tube failed to reveal tumor in only one case out of 24 tested, scanning of the tumor bed was particularly valuable, and that since P³² radioactivity was retained within the brain tissue for several days, unavoidable delay in surgery still did not invalidate the test when craniotomy could finally be performed. AMYES et al. (1955) were able to locate 14 of 15 expanding lesions, the one failure being a case of abscess. The diagnosis of intracranial neoplasm was correctly made in 46 of 50 cases by SCHNEIDER et al. (1955), but exact delineation of tumor was not emphasized. In 1958, ROBINSON and SELVERSTONE reported on a new and improved probe with a sensitive tip, which was felt to be more practicable for surface counting in areas where probe insertion might produce damage, such as the pons, medulla and optic nerves.

In summary, P³² and the miniature Geiger counter for the detection and delineation of intracerebral tumors at the time of surgery appears to be established as a method of considerable value. While the application is a limited one, it serves to reinforce the validity of this method of diagnosis.

Breast. The pioneer work of Low-Beer (1946) indicated that clinical evaluation of breast tumors might be possible with P³². Dmitrieva (1957) surprisingly found no significant difference in the radiation index from the skin area over the tumor, whether benign or malignant, in testing 16 patients, 13 with cancer, and three with benign tumors. Holan et al. (1962) reported an accuracy of 66.6 per cent in 15 patients who were given P³² orally prior to testing, and noted that the method appeared to be useful in the investigation of malignant tumors, and specifically in determining the proper time for surgical intervention following irradiation of breast cancer. More promising results were reported by Ponomareva (1959). Patients in his series were given 0.07—0.12 μc of radioactive phosphorus by mouth, and tested at two, six, and 24 hour intervals with an end-window counter applied over the tumor and compared with the normal breast. He found three false positives in normal breasts, and one false negative in a case of scirrhous carcinoma. There was an overall accuracy of 85 per cent. He stated that similar results had been obtained by Russian colleagues (Akimochkina and Ishchenko). Makletsova (1959) found three false positives in 22 patients with tumors of the breast proved to be benign mastopathy, and three false negatives in 24 supposedly benign lesions (all three scirrhous carcinoma). A somewhat different technique involving the use of P³² in the diagnosis of breast cancer was described by McFee et al. (1961). After administering 500 μc of sodium radiophosphate intravenously, a shield composed of a photosensitive emulsion coated on a vinyl plastic backing was immediately applied to the breast with the patient in a dark room. A heavy dressing of compresses and elastic Ace bandages was pressed over the plastic to obliterate all possible sources of light leak. The bandage was then left in place overnight for 18 hours, and removed in the morning, again in the dark room. The appearance of a dark area on the breast mold following development of the emulsion suggested the presence of malignant neoplasia. A thin layer of plastic wrap was applied between the skin to protect the emulsion from perspiration and adherence to the skin, and apparently adequate mold pressure to assure adequate breast coverage has been a problem. Positive results were obtained in four cancers, and ten normal breasts gave negative results. A later report by this same group, Ackerman et al. (1963) on a slightly larger number of patients indicated continued promising results. The rather divergent findings noted in the employment of the P³² test by the various investigators to date may be due to differences in technique as well as to differences in skin thickness over individual tumors, interfering with penetration of Beta radiation. So far as can be determined, testing at the time of surgery, in the tumor area as well as over lymph node distribution, has not been carried out.

Uterus and cervix. A significant increase had been noted in the chemical concentration of phosphorus in human carcinoma as compared with normal cervical tissue (Gold et al., 1950). A later study by this same group of investigators (Taymor et al., 1952), demonstrated approximately twice the uptake of P³² in the desoxyribosenucleic acid fraction of tissue removed by biopsy from 14 cervical cancers, as compared to that from five normal cervices, 48 hours after giving the isotope. After a standard dose of 2000 r roentgen-ray radiation to 11 cases of cancer of the cervix, the decrease in Beta radiation from previously given P³² was most marked in the desoxyribosenucleic acid fraction. ZACUTTI and TURCHETTI (1961) noted that radioactive phosphorus given prior to surgery appeared to fix most conspicuously in malignant tissues of resected uteri. Concentration of P³² in normal and malignant cervical tissue removed at operation after giving one microcurie intraperitoneally 24 hours previously was studied by PAPALOUCAS (1959). Isotope uptake in 20 patients before radiotherapy was approximately six times that found in six normal patients. Values were almost normal in six patients after treatment. Where the tumor showed no gross response to radiation, values were as high as before therapy. The uptake of P³² was felt to be useful in detecting early malignant change, and for assessing the results of treatment.

Clinical diagnostic studies were performed by MARCHESI et al. (1961). Using a small Geiger counter inserted into the cervical canal after injecting 1 μc/kg of P³² 12 and 14 hours prior to testing, they compared the cervical uptake to that obtained over the popliteal area. Measurements were carried out in a total of 89 women, of whom nine had erosions, six pseudoerosions, eight polyps, 12 chronic cervicitis, four ectropion, and 27 carcinoma. The were 20 normal controls. Average readings given showed distinctly higher uptake in carcinoma as compared with nonmalignant lesions, and increasingly higher values in carcinoma of greater histologic malignancy. Possible sources of error noted were: 1) the creation of a barrier between tumor and counter by cervical mucus, 2) improper angulation of the counter, producing incorrect readings, 3) insufficient penetration of the tip of the counter in ulcerating lesions, and 4) endophytic cancers too far beneath the surface to produce detectable radiation at the counter tip. Irradiation was noted to reduce P³² uptake in sensitive tumors; when this did not occur after treatment, the individual cancer could be considered radioresistent. The test was considered extremely reliable in the evaluation and diagnosis of cancer of the cervix. BECCHINI et al. (1962, 1963), observed similar results in measuring P³² uptake with externally applied Geiger counters six, 24, and 48 hours after injection of the isotope. The differentiation between benign and malignant lesions was best at 24 and 48 hours. The value of assessing results of radiation therapy were confirmed. Similar techniques as applied by KAWAI (1962) yielded excellent results, the accuracy of external counter testing being given as 78.9—95.5 per cent in the case of cancer, and 89.9—98.8 per cent in noncancer. RAGONESE (1963) found that different forms of cervical inflammation could not, however, be distinguished by comparison of radioactive phosphorus uptake.

Prostate and urinary bladder. Evaluation of the prostate was accomplished by HASKIN et al. (1961) by introducing a Geiger counter of the Selverstone type into the rectum after giving the patient 5 μc/kg of P³² intravenously 24 to 48 hours previously. The counter was guided by the finger of the operator, and readings were made by comparing different quadrants of the gland, as it was obvious that if the whole gland were involved, no differential was possible. Excellent results were obtained in the evaluation of palpable nodules, although it was felt that the type of counter was not entirely suitable, and that improvement might be possible by devising a pancake-type tip to the instrument which would be sensitive only in one plane. WOJEWSKI (1962) employed the same general technique in the examination of 15 cases of untreated prostatic cancer, three cases treated with female hormones for one-half to one year, 16 with benign prostatic hypertrophy, and eight normals. The dosage given was 0.1 microcurie of P³² intravenously, and readings were taken almost immediately after injection. A 20 per cent differential was noted in untreated cancer, but the treated group results were not described, and average readings only given. The question of overlap in readings in individual cases was not discussed. A completely negative note was struck by HAHN (1962). He tested the reliability of P^{32} in the diagnosis of prostatic cancer in 20 patients with histologically established disease, as compared to 20 with benign prostatic hyperplasia, and found that radiation differences between both types of pathology and the rectal mucosa were negligible. The test was considered unreliable for this purpose.

Radioactive phosphorus uptake in bladder tumors was studied by KAWAI (1962). He injected 10—20 μc/kg intramuscularly about 48 hours before operation, and then took readings over the bladder surfaces employing a specially made Geiger tube 4 cm in length and 5 mm in diameter. Radioactivity of seven papillomas, nine papillary carcinomas, five squamous cell carcinomas, and an adenocarcinoma was measured. Ratio of tumor to background was 2 at most with benign tumors, and greater than 2 in malignant tumors. Higher overall counts were noted with inflammation, but on ashed tissue no definite differential was noted, probably because of inadvertent mixing of samples.

Miscellaneous Tests

Skin. In an attempt to differentiate nevi, malignant melanomas, and other skin tumors, BAUER and STEFFEN (1955), administered 100 to 150 μc of P^{32} intravenously and tested various skin lesions by determining radiation directly from the surface with an end-window Geiger tube one to three hours later. Seventy patients, six with malignant melanoma, 10 with squamous cell carcinomas, nine with basal cell carcinomas, and the remainder with a variety of benign lesions, were evaluated. Malignant melanomas were differentiated from benign nevi by significantly increased concentrations of P^{32}, but basal and squamous cell carcinomas demonstrated no diagnostic level of uptake over benign skin lesions. BRAUER et al. (1960) employing essentially the same technique in 49 patients, found 20 per cent false positive, and 37 per cent false negative results. The overlap of results was felt to render the test unreliable for the differentiation of melanomas from other skin tumors. Increased uptake of P^{32} in melanomas was used as a means of establishing the extent of invasion of melanoma by LAZAROA-IKONOPISOV (1960), who described a method of mapping the surface and testing for radiation.

Pharynx and Larynx. The possible value of P^{32} in the diagnosis of tumors of the larynx and pharynx was evaluated by FILIPPI et al. (1960), using open-end Geiger counters on fresh or acid calcinated tissue removed 24 to 48 hours after the oral administration of 500 to 1,000 μc of P^{32}. In 12 patients who had total laryngectomy for squamous cell carcinoma, five with squamous cell carcinoma of the tonsil and pharynx, and two with reticulum cell sarcoma of the tonsil, tumor and normal tissue was removed separately, dissolved by acid calcination and made up to a standard volume. Counts taken over the dissolved tissue were expressed as counts per minute per milligram of fresh tissue. Radioautographs were also made from samples of the same tissue, and counting rates were taken in vivo in 20 patients with tumors of the larynx, pharynx or oral cavity (squamous cell carcinoma or reticulum cell sarcoma). In 13 with palpable lymph nodes, seven patients with superficial nodes were scanned, and background counts compared. In all cases, the homogenized tissue, scans, in vivo scans, and radioautographs showed increased radiation of a significant degree in the case of malignant neoplasia as compared to normal tissue. Superficial lymph nodes scanned in vivo produced only slightly greater radiation than background tissue, which could not be considered significant. The indeterminate differences obtained in superficial nodes was felt to be due probably to difficulty in penetration of the overlying skin; it was postulated, but not demonstrated, that scanning of nodes at surgery might be of more value. The conclusion was drawn that scanning over tumors may give additional information in diagnosis but is supplementary to other means. In a preliminary study, PALLESTRINI (1961), using the radioautograph strip film technique in 20 patients, demonstrated selectively higher uptake of P^{32} in laryngeal tumors and metastatic lymph nodes in 20 patients. NAGATANI (1960), after first demonstrating increased uptake of radioactive phosphorus by radioautographic studies, carried out in vivo scanning in 78 patients, some of whom had tumors of the mouth, pharynx or nose. Counts were taken one, three and six hours after administration of P^{32}. Eight cases with malignant tumors were correctly diagnosed by this technique prior to surgery and histologic examination.

Bone. Autoradiographic studies in rats and dogs (GUILMET et al., 1963), demonstrated localization of most of the P^{32} in the medullary portions, which were removed to obtain accurate images of the cortex. In metastatic cancer of the breast, bone lesions may be demonstrated by radioautographs, most of the P^{32} accumulating in regenerating bone surrounding the metastasis (KAPLAN et al., 1959). JUSTUS (1961) found a difference in radiation in bone tumors of 100 to 200 per cent in malignant tumors as compared to controls, and felt that radioactive phosphorus might have some clinial value in diagnosis.

Pleural effusions. The use of P^{32} in the differential diagnosis of pleural effusions was investigated by BAUER et al. (1954). Pleural fluid and specimens of whole blood were removed 24 hours after giving 100 to 200 microcuries of P^{32}, and aliquots were tested for radiation in a shielded lead holder, using Geiger counters with thin end-window mica tubes. Sediment obtained by centrifugation was examined histologically. Two groups of patients consisting of nonmalignant diseases, the first with noninflammatory, and the second with inflammatory disease, were compared with a third group of 16 patients with malignancy and pleural effusions, of which 7 were metastatic and 9 primary cancer of the lung. Radiation measured definitely higher in the malignant neoplasia group, although there was some individual overlap. Future usefulness in distinguishing between benign and malignant pleural effusions was expected. Repeating essentially the same study, TAYLOR et al. (1958), tested 50 patients with various inflammatory and malignant effusions. They found a statistically significant difference between the acute tuberculous effusions and those secondary to malignant neoplasia, although there was still some overlap. They noted no difference between effusions in the malignant group and those due to chronic tuberculosis, nonspecific inflammations, pulmonary infarction, Meig's syndrome and spontaneous pneumothorax. Radiotherapy appeared to lower radiation values in the malignant neoplasia group, but the results were considered to be too equivocal in general for the test to be of value in the differential diagnosis of possibly malignant effusions.

Summary

The use of radioactive phosphorus in the differential diagnosis of malignant and benign lesions outside the gastrointestinal tract, while disappointing in some areas, showed considerable promise in others. The limitations in the method are the relatively low penetrating ability of the Beta radiation concentrated in malignant neoplastic tissue, as well as the necessity to position the Geiger counter (or other scanning means) rather exactly to obtain definite differential readings between malignant and normal tissue. Where the tumor is accessible, and not covered with an excessively thick normal layer containing its own background radiation, excellent contrast readings are possible, with clear-cut diagnoses later borne out by histologic study. Where good counter position is impossible, or covering normal tissue too impenetrable, equivocal radiation is recorded, and no diagnosis is possible. These factors must be carefully considered when weighing the value of a positive or negative finding in the individual patient, and constitute just cause for considering the P^{32} test as a part of the diagnostic picture only, rather than conclusive in the majority of instances.

References

ACKERMAN, N. B., A. S. MCFEE, J. A. BLUM, E. L. MAKOWSKI, and O. H. WANGENSTEEN: Multiple-organ cancer screening with diagnostic radioautographic procedures. J. Amer. med. Ass. 183, 36 (1963).

AMYES, E. W., P. H. DEEB, P. J. VOGEL, and R. M. ADAMS: Determining the site of brain tumors; the use of radioactive iodine and phosphorus. California Med. 82, 167 (1955).

BAUER, F. K., and C. G. STEFFEN: Radioactive phosphorus in the diagnosis of skin tumors; differentiation of nevi, malignant melanomas, and other skin tumors. J. Amer. med. Ass. 158, 563 (1955).

BAUER, R. E., I. H. MOSS, and A. D. RICHARDSON: A study of radioactive phosphorus activities in pleural effusions. Cancer 7, 852 (1954).

BECCHINI, M. F., A. ZACUTTI, and G. TURCHETTI: The use of radioactive phosphorus (P^{32}) in the diagnosis of cancer of the uterine cervix. Minerva Nucl. 6, 23 (1962).

BETTMAN, J. W., and V. FELLOWS: Radioactive phosphorus as a diagnostic aid in ophthalmology. Arch. Ophth. 51, 171 (1954).

BRAUER, E. W., A. W. KOPF, V. H. WITTEN, M. BERMAN, and V. G. CAVE: Radioactive phosphorus in the in vivo diagnosis of melanoma of the skin. J. Amer. med. Ass. 172, 1753 (1960).

CARMICHAEL, P. L., and I. H. LEOPOLD: The radioactive phosphorus test in ophthalmology. Amer. J. Ophth. 49, 484 (1960).

DMITRIEVA, P. E.: Use of radioactive phosphorus for diagnosis of breast cancer. Eksp. Khir. 2, 37 (1957).

DONN, A., and J. W. MCTIGUE: The radioactive phosphorus uptake test for malignant melanoma of the eye; a study of forty consecutive cases of suspected intraocular malignant melanomas. Arch. Ophth. 57, 668 (1957).

DUNPHY, E. B., J. L. DOWLING, JR., and A. SCOTT: Experience with radioactive phophorus in tumor detection. Arch. Ophth. 57, 485 (1957).

EISENBERG, J. J., I. H. LEOPOLD, and D. SKLAROFF: Use of radioactive phosphorus in detection of intraocular neoplasms. Arch. Ophth. 51, 633 (1954).

ERF, L. A., and J. H. LAWRENCE: Phosphorus metabolism in neoplastic tissue. Proc. Soc. exp. Biol. Med. 46, 694 (1941).

FILIPPI, P., O. FERRINI, and G. CREMONESI: Radiophosphorus (P^{32}) in the diagnosis of tumors of the larynx and pharynx. Acta otolaryng. 52, 221 (1960).

GARRITY, R. W., and L. W. MATTHEWS: Radioactive phosphorus in management of brain tumors. Neurology 4, 929 (1954).

GOLD, N., M. TAYMOR, and S. STURGIS: Studies on labelled phosphorus in cancer of the cervix. Surgical Forum (1951), p. 544.

GOLDBERG, B., D. TABOWITZ, G. B. KARA, S. ZAVELL, and R. ESPIRITU: The use of P^{32} in the diagnosis of ocular tumors. 1. A clinical report of 125 cases. Arch. Ophth. 65, 196 (1961).

GUILMET, D., M. DUBRASQUET, and S. BONFILS: Bone (P^{32}) autoradiography method. Technic. Experimental importance. Rev. franç. Étud. Clin. Biol. 8, 717 (1963).

HAHN, M.: Isotopes in the diagnosis of prostatic carcinoma. Rozhl. Chir. 41, 680 (1962).

HASKIN, M. E., M. L. WAGNER, M. IVKER, and B. P. WIDMANN: The early diagnosis of prostatic malignancy by the use of P^{32}. Amer. J. Roentgenol. 85, 99 (1961).

HOLAN, T., M. MICLUTEA, S. COMES, T. VASCULESCU, and H. D'ANDRE: Unsere Erfahrungen mit radioaktivem Phosphor in der Diagnostik des Brustkrebses. Radiobiol. Radiother. 3, 215 (1962).

JONES, H. B., I. L. CHAIKOFF, and J. H. LAWRENCE: Phosphorus metabolism of neoplastic tissues (mammary carcinoma, lymphoma, lymphosarcoma) as indicated by radioactive phosphorus. Amer. J. Cancer 40, 243 (1940).

JUSTUS, G.: P^{32} in the diagnosis of bone tumors. Zbl. Chir. 86, 298 (1961).

KAPLAN, E., J. MIREE, JR., E. HIRSH, and T. FIELDS: Autoradiographic localization of P^{32} phosphate in metastatic carcinoma of the breast to bone. Internat. J. appl. Radiat. 5, 94 (1959).

KAWAI, I.: Fundamental and clinical studies on the diagnosis of uterine cancer using P^{32}. J. Osaka City Med. Cent. 11, 63 (1962).

KAWAI, T.: On the radioisotope treatment in the field of urology. V. Diagnostic application of radioactive phosphorus to the bladder tumor. Jap. J. Urol. 53, 681 (1962).

KENNEDY, R. J., O. GLASSER, and P. KAZDAN: Use of radioactive phosphorus in detection of intraocular tumors. Clevel. Clin. Quart. 21, 133 (1954).

KROHMER, J. S., C. I. THOMAS, J. P. STORAASLI, and H. L. FRIEDELL: Detection of intraocular tumors with the use of radioactive phosphorus. Radiology 61, 916 (1953).

LAZAROV-IKONOPISOV, R.: Malignant melanoma of the skin. 1. A mode to establish the extent of the subclinical permeation and infiltration of malignant melanomas of the skin by means of the P^{32} uptake test. Neoplasma 7, 377 (1960).

LIDGETT, K., K. H. CLARKE, and H. A. VAN DEN BRENK: Studies of radioactive phosphorus uptake in the diagnosis of intraocular and cutaneous melanomata. Aust. N. Z. J. Surg. 29, 149 (1959).

LOW-BEER, B. V. A., H. G. BELL, H. J. McCORKLE, and R. S. STONE: Measurement of radioactive phosphorus in breast tumors in situ; a possible diagnostic procedure. Radiology 47, 492 (1946).

MAKLETSOVA, N. P.: Radioactive phosphorus (P^{32}) in differential diagnosis of certain superficially placed malignant and benign neoplasms. Vop. Onkol. 5, 702 (1959).

MARCHESI, F., M. MANESCHI, and E. CITTADINI: The differential diagnosis of carcinoma of the uterine cervix using P^{32}. Panminerva Med. 3, 508 (1961).

MARINELLI, L. D., and B. GOLDSCHMIDT: The concentration of P^{32} in some superficial tissues of living patients. Radiology 39, 454 (1942).

MARSHAK, A.: Uptake of radioactive phosphorus by nuclei of liver and tumors. Science 92, 460 (1940).

McFEE, A. S., N. B. ACKERMAN, and O. H. WANGENSTEEN: In vivo P^{32} radioautography in detecting breast cancer. Proc. Soc. exp. Biol. Med. 107, 847 (1961).

MORLEY, T. P., and G. JEFFERSON: Use of radioactive phosphorus in mapping brain tumors at operation. Brit. med. J. 2, 575 (1952).

NAGATANI, K.: Diagnostic applications of radiophosphorus for malignant neoplasms in otorhinolaryngology. Acta Otolaryng. 52, 501 (1960).

O'ROURKE, J., and E. COLLINS: P^{32} localization of malignant melanoma of the posterior chorid. Arch. Ophth. 63, 801 (1960).

O'ROURKE, J. F., H. PATTON, and R. BRADLEY: Fundamental limitations of radiophosphorus counting methods used for detection of intraocular neoplasm. Arch. Ophth. 57, 730 (1957).

PALLESTRINI, E., and P. FILIPPI: Autoradiographic research on the fixation of P^{32} by laryngeal carcinomas and their satellite lymph nodes. Acta otolaryng. 53, 271 (1961).

PAPALOUCAS, A. C.: P^{32} uptake in treated and untreated cervix uteri carcinoma. Nucl. Med. 1, 62 (1959).

PAROLI, G., A. ZACUTTI, G. TURCHETTI, and M. F. BECCHINI: The diagnosis of female genital tumors with radioactive isotopes. Minerva Nucl. 7, 135 (1963).

PONOMAREVA, G. M.: The diagnosis of cancer of the breast with the aid of radioactive phosphorus. Vop. Onkol. 5, 706 (1959).

RAGONESE, P.: Value of the P^{32} phosphorus uptake test in inflammation of cervix. Panminerva Med. 5, 312 (1963).

RIORDAN, D. J., P. A. BROWNE, and P. J. KENNY: Radioactive phosphorus in the diagnosis of ocular malignancy. J. Irish med. Ass. 44, 82 (1959).

ROBINSON, C. V., and B. SELVERSTONE: Localization of brain tumors at operation with radioactive phosphorus; an improved technique using a proportional counter. J. Neurosurg. 15, 76 (1958).

SCHNEIDER, R. C., H. PANTEK, D. G. FREEMAN, and R. G. FARRIS: Radioactive phosphorus in the localization of the brain tumors. J. Mich. med. Soc. 54, 434 (1955).

SELVERSTONE, B., and J. C. WHITE: Evaluation of the radioactive mapping technic in the surgery of brain tumors. Ann. Surg. 134, 387 (1951).

—, A. K. SOLOMON, and W. H. SWEET: Location of brain tumors by means of radioactive phosphorus. J. Amer. med. Ass. 140, 277 (1949).

SHAPIRO, I.: Radioactive phosphorus in differential diagnosis of ocular tumors. Arch. Ophth. 57, 14 (1957).

TAYLOR, A. J., J. D. PEARSON, and N. VEALL: The use of radioactive phosphorus in the diagnosis of pleural effusions. Brit. J. Tuberc. 52, 281 (1958).

TAYMOR, M. L., H. GOLD, S. H. STURGIS, J. V. MEIGS, and J. MacMILLAN: Effect of irradiation upon the uptake of labelled phosphorus in human carcinoma of the cervix. Cancer 5, 469 (1952).

THOMAS, C. I., J. S. KROHMER, and J. P. STORAASLI: Detection of intraocular tumors with radioactive phosphorus. Arch. Ophth. 47, 276 (1952).

—, M. S. BOVINGTON, and J. S. KROHMER: Uptake of radioactive phosphorus in experimental tumors. 1. Comparison of radioactivity in neoplastic and normal ocular tissue. Cancer Res. 16, 796 (1956).

—, J. P. STORAASLI, and H. L. FRIEDELL: Radioactive phosphorus in the detection of intraocular neoplasms. Amer. J. Roentgenol. 95, 935 (1965).

—, M. S. BOVINGTON, J. S. KROHMER, J. W. MacINTYRE, and J. P. STORAASLI: Uptake of radioactive phosphorus in experimental tumors. II. Effect of vascularity on radioactive uptake in neoplastic, inflammatory and normal ocular tissue. Cancer Res. 17, 1091 (1957).

TUDWAY, R. C., and A. PALIN: The use of radioactive phosphorus for the detection of intraocular tumors. Brit. J. Radiol. 31, 549 (1958).

WOJEWSKI, A.: The evaluation of methods applied in diagnosis of prostatic cancer. Urol. In. 14, 140 (1962).

ZACUTTI, A., and G. TURCHETTI: Radioautographical study of the fixation of radioactive phosphorus in benign and malignant tumors of the uterus. Arch. Ostet. Ginec. 66, 303 (1961).

Chapter II

Radioactive Phosphorus (P³²) in the Diagnosis of Gastrointestinal Cancer

Clinical use of the radioactive phosphorus (P³²) test naturally expanded first in the more accessible regions of the body. Areas such as the eye, skin and breast could be carefully evaluated for relatively long periods and, if necessary, repeatedly with little discomfort to the patient. The difficulty of instrumentation in the gastrointestinal tract presented a natural barrier to clinical evaluation. During the period of early delay, a basic study (SCHULMAN et al., 1949) on phosphorus turnover in human gastric carcinoma, shed some light on the possible usefulness of the test in the stomach. Following subtotal gastrectomy in three patients with nonmalignant disease and six with various gross types of cancer, who had been given 1 μc. of P³² per pound of body weight 36 hours prior to surgery, mucosa was stripped from the stomach wall in all cases, and samples taken of tumor mass. Analysis showed that the phosphorus content of mucosa from stomach with either benign or malignant disease, was essentially the same. The phosphorus content of gastric mucosa, whether associated with benign or malignant disease, and cancerous tissue, was also the same, but the rate of phosphorus turnover was 47 per cent higher in gastric carcinoma than in other gastric tissue. There was a 124 per cent increase in the turnover of protein phosphorus and a 45 per cent increase in the turnover of lipid phosphorus, but the acid-soluble phosphorus turnover remained unchanged. The increased activity of tumor phosphorus in the stomach demonstrated the possible usefulness of the P³² test in differential diagnosis of gastric lesions.

Noting that little attention had been paid to the possibility of employing the P³² test in the diagnosis of malignant neoplasia of the gastrointestinal tract, NAKAYAMA (1956) presented a comprehensive study of the method in lesions of the esophagus, stomach and rectum. Aliquots of cancerous and normal tissues from patients with cancer of the stomach, esophagus and rectum who had previously received radioactive phosphorus were evaluated as to counting ratios. The results confirmed the uptake of P³² by malignant tissues in markedly increased ratio, and definitely indicated more active metabolism of phosphorus in malignant neoplasia than in normal tissue. Couting rates, when measured with the Geiger counter over chronic inflammatory areas, remained less than 1.3 times the normal count, while macro- and microradioautography demonstrated the selective uptake of P³² over surrounding normal tissue. In clinical application, 5 to 10 μc/kg; or a total of 300 to 500 μc., was administered intramuscularly 6 to 48 hours prior to testing. The Geiger counter was inserted in the end of catheters for internal scanning, a somewhat smaller counter being used for the stomach than for the esophagus. In the esophagus, counting was carried out by passing the counter through the lumen of the esophagoscope, and in the stomach the instrument was attached to inflated balloons and applied to the wall under fluoroscopic control. Rectal testing apparently was carried out through the sigmoidoscope. Esophageal tests were made in 75 cases, of which 58 were malignant lesions, including 39 esophageal cancers, 18 cardiac cancers, and one esophageal sarcoma. Accuracy was 100 per cent. In the stomach, there was 97 per cent accuracy in 61 malignant lesions, and 100 per cent accuracy in 29 cases with benign lesions. Seventy control stomachs all gave normal background counts. Five of six cases of rectal cancer gave diagnostic readings. The results of the study were reported as showing that the P³² test provided specific information not otherwise obtainable, and when added to conventional means, such as x-ray, endoscopic study and cytology, allowed a much more accurate diagnosis of cancer in the alimentary tract.

ACKERMAN et al. in 1960 reported on a different technique for diagnosing cancer of the stomach, employing P³² as the tracer element. They had invited NAKAYAMA to demonstrate his method at the University of Minnesota, and duplicated his equipment, but with limited success. Succeeding studies with in vitro radioautography of gastric lesions, lymphosarcomas as well as adenocarcinomas, convinced them of the selective uptake of radioactive phosphorus in malignant tissue. They devised a latex balloon coated with a photosensitive emulsion on the outside, inverted it, and attached it to the open end of a four-hole nasogastric tube. Following the oral ingestion of P³², the deflated balloon was passed through the nose into the stomach in a photographic dark room. The balloon was then inflated with 600 to 900 ml. of air and kept in place for four hours. Seconal and probanthine were given to control restlessness and secretions. When the time was up, the patient was returned to the dark room, and the balloon deflated and removed. Injection of 150 cc. of the developing solution into the balloon completed the study, which was judged positive when a darkened area appeared denoting selective absorption of P³² in a cancer which had been adjacent to that area of the balloon. Seven patients with gastric cancer all gave positive results. In ten benign gastric ulcers three gave false positive results. Three patients with normal stomachs all gave negative results, five with benign polyps were negative, and eleven with pernicious anemia, but without evidence of gastric lesions, were negative, as were 19 with achlorhydria but otherwise normal stomachs. Two with hypertrophic rugae also failed to discolor the balloon. Although the false positive results in benign gastric ulcer were deemed puzzling, the method of testing was thought to be very promising. A later report by the same group (ACKERMAN et al., 1962) presented refinements in technique of the same method. Intravenous administration of P³², rather than oral, was advocated because of the possibility of residual radioactivity in the stomach, and the isotope was given 12 to 18 hours before testing. Improved and more sensitive emulsions which permitted a shortening of the exposure time, preshaped balloons for more accurate approximation to the stomach wall, and a double lumen tube for removing gastric secretions were used to promote greater accuracy and limit patient discomfort. The balloon was inverted on a glass tube and immersed in the developing solution in the dark room. Improved results were claimed with these changes, and several examples of positive tests illustrated, but no further statistics given. In a later report, this same group (ACKERMAN et al., 1963) stated that 25 of 33 gastric carcinomas had been correctly diagnosed by this method, and that eight false positive readings had been obtained among 64 benign gastric ulcers studied. Rather low sensitivity of the photographic emulsion of the balloon was given as the main limiting factor for greater accuracy.

Radioactive phosphorus (P^{32}) and NaI^{131} were evaluated by OTTO et al. (1964), who concluded that there was no selective uptake of NaI^{131} in 12 gastric carcinomas studied preoperatively by scintiscans, but that the use of P^{32} might be promising. Four hundred microcuries of P^{32} were given intravenously 18 to 24 hours prior to examination. Lesions were scanned with a solid state detector rather than a Geiger counter. This instrument was attached by subminiature coaxial cables to a rate meter, a spring loaded deflector and thin hollow polyethylene tubing were placed adjacent to the cable and the entire apparatus encased in silicon rubber covering, leaving an outlet for the polyethylene tubing so that air could be pumped into the stomach. The stomach was inflated with air after passing the apparatus, and using image intensified fluoroscopy, the detector was placed first against normal mucosa and then against the lesion as determined by previous roentgen studies. In some cases (number not given) the detector was positioned under direct vision during gastroscopy. Positioning was usually accomplished, however, by rotation of the patient in the erect, prone and supine positions and by the use of the deflector. Time given for the whole procedure was 10 to 15 minutes. In 16 cases of carcinoma, accuracy was 93.8 per cent with a median uptake ratio of 1.8 : 1, and in 37 cases of benign lesions, including peptic ulcer, pyloric hypertrophy, leiomyoma and adenoma, 97.3 per cent with a median uptake ratio of 1 : 1 to 1.25 : 1. One false positive and one false negative were encountered. The case demonstrations used in the report showed rather large lesions; use of the detector for small polyps or general stomach scanning was not mentioned.

In 1963, VOSNYUK described the use of a small Geiger counter seated in the end of a flexible gastroscope for the detection of gastric malignancy in patients previously given P^{32}, but details as to his testing and accuracy were not available. He also attempted to determine radioactivity of gastric washings in such patients with little success. STEIN et al. (1965) reported poor results with the sensitized balloon technique of ACKERMAN. In comparing this method with radiography and cytology, they found that in 12 patients with histologically proven cancer of the stomach, 11 were diagnosed as definitely or probably malignant by roentgenoscopy, and seven by cytology. Balloon radioautography was positive in four only. In 26 patients with histologically verified benign lesions, 22 were reported as benign or probably benign on roentgenological examination, and only three as probably malignant. Diagnosis by exfoliative cytology was benign in all but one, which yielded a false positive. Balloon radioautography indicated that 23 were benign, and four showed discoloration of the balloon, rated as false positives. In a second group of 25 benign lesions not proven histologically, but in which the diagnosis seemed definite on follow-up, the x-ray was rated as definitely or probably benign in 24, and indeterminate in one, later classified as benign. Exfoliative cytology was negative for malignant cells in all 25, but balloon radioautography showed two false positives. The authors pointed out that with the large number of negative studies, a false opinion of the worth of this test might be gained overall, since if no sensitive coating were applied to the balloon, the test would still have yielded an "accuracy" of 81.3 per cent in the 64 cases studied. In discussing their inability to reproduce ACKERMANs results, especially in carcinoma, they noted that artifacts due to light or previous radiation were probably responsible for some false positives, that it was virtually impossible to confirm the exact position of the balloon within the stomach, even though close approximation to the mucosa was necessary, and that the presence of residue within the stomach might be sufficient to prevent close contact of the balloon with the lesion. The final conclusion was that while balloon radioautography was probably too inaccurate for further study, the use of P^{32} as a test for gastric cancer might still be valuable if a solid state detector or Geiger counter could be applied closely to the lesion.

The use of radioactive phosphorus (P^{32}) in the diagnosis of gastrointestinal lesions, since the original work of NAKAYAMA in 1956, appears to have been rather limited. All workers apparently agree that the method may be promising, although techniques differ. Most of the published cases, yielding positive results, are those of rather large gastric lesions, which, while demonstrating the feasibility of the method, leave some doubt as to its usefulness in the differential diagnosis of small questionable areas in the stomach, or in scanning the stomach as a whole. The esophagus and rectum, except for NAKAYAMA's report, are neglected. It is apparent from all publications that close approximation to the mucosa, whether by photosensitive balloon, solid state detector or Geiger counter is essential to success. All types of apparatus have drawbacks, largely based on the difficulty of accurate placement within the accessible regions of the gastrointestinal tract.

Original Research

Our own studies were initiated by NAKAYAMA's 1956 report. A start was made in 1957 with the procurement of a number of miniature Geiger counters, followed by early attempts to insert them into the stomach for the purpose of detecting radioactivity of lesions in patients previously given P^{32}. One of the major problems encountered was the determination of the proper size of counter to be used. First experiments were carried out with the smallest type available, but these were found to be too insensitive to give good differential readings, even when applied to freshly removed tissues in the operating room. There were obvious limitations to the size of instrument that could be employed in the esophagus which would still go through the cardioesophageal junction. The conclusion was eventually reached that the largest counter fulfilling this criterion would probably be best. In early studies, the esophagus, stomach, and rectum were all tested with the same Geiger tube, although it was apparent that it was possible to use a much larger instrument in the rectum, and that this might be preferable. Permanent graph recordings, rather than single maximum deviations, appeared to have value from the first, especially in the esophagus and rectum, where lesions could be located and recorded as a function of depth and so be more precisely delineated.

An extensive background in endoscopy of the gastrointestinal tract in the demonstration of malignant neoplasia made it appear doubtful that any type of counter could be applied to all areas of the gastric mucosa under direct vision. Although it would obviously be easy to scan large stomach lesions, and compare radioactivity with that in the esophagus or other normal tissue, it was anticipated that difficulty would be encountered with polyps or questionable localized infiltrations of the gastric wall. The visualization of lesions through open-tube endoscopes such as the esophagoscope and sigmoidoscope presented fewer difficulties, and the accurate approximation of a Geiger counter to all areas, including those involved by tumors, more likely of success. In the stomach, only a small part of the fundus can be visualized through the esophagoscope, and all folds are pressed together. The stomach must be inflated for good vision, and the flexible gastroscope (whether lens or fiberoptic in type) must be employed, since the old rigid instrument is too dangerous for routine use. Considerable gastroscopic experience is necessary for recognition and proper evaluation of stomach findings with the flexible gastroscope, and the inflated stomach presents a rather large area, with relatively great distances between the walls. All things considered, the accurate placing of a small Geiger counter on any small area within the inflated stomach, even with the aid of an attached balloon and under fluoroscopy, must be regarded with some skepticism. Even from the first, therefore, it was recognized that the P³² test would be much more difficult to perform accurately in the stomach than in the esophagus and rectum.

Other main points of consideration during early studies were those of convenience, simplicity, cost, and the safety of the patient. Any test so complicated as to limit its use will probably be of limited value in the overall assault on gastrointestinal cancer, however intriguing some of its results may appear. Although some might feel that the extensive employment of endoscopy in gastrointestinal cancer diagnosis is unnecessary, our experience has been that if all available types of endoscopy were used earlier and more widely, a considerable improvement in mortality from cancer of the esophagus, stomach and rectum would result. Endoscopy is used routinely on our service. The addition of the P³² test did not, therefore, present any difficulties based on general considerations of danger, inconvenience or cost, since the added expense of radioactive phosphorus and the apparatus (which may be used for hundreds of tests) is minimal.

Despite the rather considerable voltage necessary to perform the test, safety precautions involved in the insulation of the counter and connections have proved no problem. There have been no endoscopic accidents, and no instances of shocking or current leak in over 250 tests. Similarly, radiation exposure probably is within safety limits. The administration of 500 μc. delivers a radiation dose of about 15 r. to the bone marrow, and although this amount is at least 100 times greater than that delivered from routine chest x-rays, it is less than the 30 r. per year which has been established as the permissible limit for radiation workers. It may be possible to reduce this amount in the future with improved equipment. Nevertheless, radiation is a consideration if the test is repeated in the same patient, when time intervals, age of the individual, and chance of eventual cure must all be taken into account.

During the past nine years, considerable experience has accumulated on our service in the use of radioactive phosphorus in the detection of gastrointestinal cancer. As expected, the stomach has presented many more difficulties in testing than the esophagus and the rectum. Specific situations in all three organs have given rise to particular problems, and subsequent attempts at solution which will be presented in detail.

1. Method in General

The P³² test may be used to evaluate lesions noted by previous x-ray examination, or as a routine survey of the mucosa of esophagus, stomach, or rectum. Testing is entirely in practicable in out- as well as inpatients. The tracer dose of 500 μc. of radioactive phosphorus is administered intravenously 18 to 20 hours before examination, either in the outpatient clinic, or on the ward for patients who are ordinarily examined the following morning in the endoscopy room. When the patient has been prepared for endoscopic examination in the usual manner, isotope service is notified, and the equipment brought to the clinic on a small cart by the technician, who checks the Geiger counters and records the tests. When the examination is complete, the recording is made part of a permanent file on the isotope section. Duplicates may be made for the files of the attending physician, or for the patient's chart.

2. Equipment in General

Geiger counter tubes. Counters employed in this study are of two types: 1) A Geiger tube 25×5 mm. in dimension, with a sensitive area of 5×10 mm. has been used most extensively in the esophagus, stomach and rectum. This tube has an intrinsic efficiency of about 50 per cent for detecting P^{32} Beta particles and gives a counting rate of about 5500 counts per minute when placed above a solution of P^{32} with a concentration of 1 μc. per milliliter. For testing the esophagus and rectum, the counter is incorporated in the tip of a fairly rigid Teflon plastic tube approximately 60 cm. long, which allows the counter to project about 2 cm. beyond the open tip of the esophagoscope and sigmoidoscope, with the proximal end of the plastic tube under control of the operator. The same counter has also been inserted in the end of a controllable tip gastroscope, just distal to the objective window and lamp, for evaluation of the stomach. In both types of installation, the Geiger tube is carefully protected by a thin Teflon covering 0.075 mm. in thickness. The other type of counter, 2) is a standard miniaturized end mica window tube, ordinarily used for eye counting, which has been modified by removing 90 per cent of the wall thickness

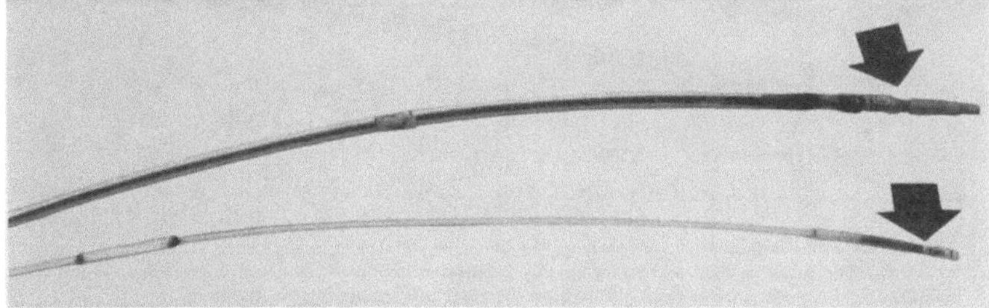

Fig. 1. Comparative sizes of two types of Teflon probes and Geiger counter tubes employed in this study are shown. The larger probe and counter are used to test the rectum only; the smaller may be used in esophagus or rectum. Arrows denote location of counters

along .6 inches of the tube. Overall dimensions are $2.15 \times .255$ in., and efficiency approximately 80 per cent for detecting Beta particles. This tube is incorporated in the tip of a larger and somewhat more flexible Teflon tube for insertion through the sigmoidoscope, and is protected by the same type of thin Teflon covering 0.075 mm. thick. This probe has been used exclusively in scanning the rectum. Its dimensions are such as to preclude its employment in either the esophagus or stomach (Fig. 1).

Recording equipment. A Baird Atomic Ratemeter, model 432 A, modified for use with the special gastroscope, esophageal and sigmoidoscopic probes, as well as for use with a Texas Instrument Recorder, is employed to obtain radiation values, which are then recorded in the form of a graph for each patient.

Criteria for evaluation. Accuracy of the testing was based, in the case of proven neoplasia, upon the correlation of a histologically proven lesion with at least a 30 per cent increase in counting rate over that seen in the adjacent control area (ordinarily 100 to 200 counts per minute [c.p.m.]). The background counting rate with the tube removed from the stomach was 10 c.p.m. In nonneoplastic disease, the lack of an increase in counting rate was correlated with negative biopsy specimens and a sufficiently long observation time to confirm negative evaluation. In positive tests, the counting rate over the tumor area ranged from 30 to 200 per cent higher than that over the control area.

3. Testing in Individual Organs

a) Esophagus

Methods. Esophagoscopy is carried out in routine fashion except for the addition of the P³² technique. The patient is fasted from the previous midnight, Meperidine, 50 mg., and sodium pentobarbital, 100 mg. are given intramuscularly as sedation, and pontocaine, 2 per cent, applied to the mouth and pharynx either as spray or gargle for local anesthesia. The Eder-Hufford esophagoscope, 53 cm. length model, with 9.5 mm. inside diameter, is passed under direct vision, into the upper stomach if possible. The Geiger probe is then inserted, and passed to the point that two centimeters are projecting beyond the tip of the esophagoscope (Fig. 2). Initial readings

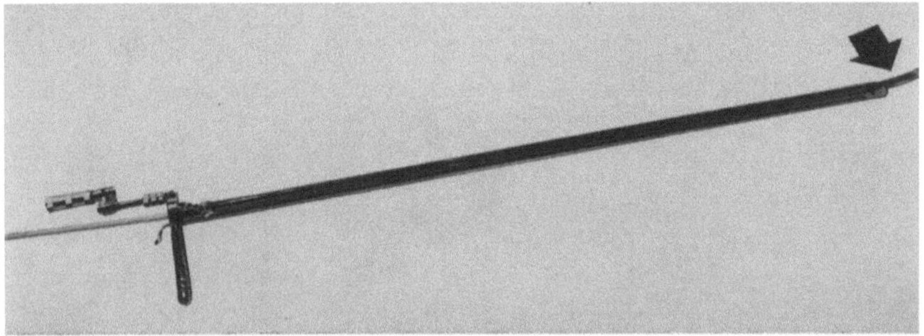

Fig. 2. The smaller probe is shown inserted through the lumen of the esophagoscope with Geiger counter projecting, noted by arrow. The probe is held stationary in this position as the esophagoscope is retracted. Distances from the incisor teeth are read on the outer instrument in centimeters

are taken and recorded on the moving graph, being careful to obtain a sustained reading with as little variation as possible, which may take from 15 to 20 seconds initially. When the first stable count has been recorded, the technician notes the location of the reading on the graph and tells the operator to proceed. He retracts the esophagoscope, with the enclosed probe firmly held at the same level, for a distance of approximately two centimeters, and a new reading is recorded until stable. This process is continued the length of the esophagus, up to approximately the 20 centimeter level, when the probe is removed. If a lesion is encountered during the passage of the esophagoscope, it may be possible to advance the tip of the instrument beyond this point if there is not complete obstruction. If possible, this should be carried out, since it is important to determine the extent of the lesion where feasible. Good approximation of the Geiger counter to the tumor or other lesions is also best obtained with the tip of the esophagoscope as close as possible to the affected area. With a visible lesion, readings are obtained in the previously described manner, at least to the level where good background readings are obtained in the case of a positive test. If radiation differential is not clearly diagnostic, the process may be repeated in an attempt to clarify the situation. When a positive reading is obtained, the counter is removed, the esophagoscope again passed to the involved area, and biopsies taken. Even when gross tumor is not identified, and the mucosa appears relatively normal, a diagnostic radiation differential is the signal for several biopsies within the area signified by the graph. True stricture of the esophagus is difficult to test, and vigorous efforts should be made to pass the probe tip into the residual opening in the esophagus, since this is the only manner in which accurate testing can be carried out. Large exophytic tumors are simple to evaluate. The counter usually comes into direct approximation with the surface of the lesion with minimal manipulation. Inflammatory lesions are also easy to test as a rule. A certain amount of fluctuation of recording will ordinarily be noted, probably due to pressure changes within the thorax during respiration, and consequent differences in distance between the Geiger counter and esophageal mucosa. These may be minimized by proper adjustment of the recorder which will also compensate for a large malignant lesion. The high differentials given by recordings are read following the conclusion of the procedure, and decision as to whether a lesion is malignant or benign may be made on the spot, final correlation being obtained as a result of the biopsy or surgical resection, when a diagnostic increase in radiation is noted.

b) Results

Primary cancer of esophagus. In 47 patients, initial testing was positive in 45, and negative in two, for an accuracy of 95.7 per cent. The two false negatives were obtained early in the study in patients with considerable esophageal narrowing which rendered proper apposition of the counter difficult. No false negative tests have been encountered in this category of patients in the past four years. Follow-up studies in several patients have been accurate in all instances (vide infra, case reports).

Table 1. *Cancer of Stomach (Involving Esophagus)*

Gross Esophagoscopic Findings (Groups)	Total Patients	Biopsy Findings			Results of P^{32} Test	
		None	Cancer	Negative	Positive	Negative
1. Tumor masses seen	4		4	0	4	0
2. Submucosal nodules	5	1	3	1	4	0
3. No gross involvement	13	3	3	7	11	2

Gastric cancer invading esophagus. Tests in 23 patients have yielded two false negative results, for an accuracy of 91.3 per cent. The esophagoscopic findings in this group of patients may be classified as 1) four patients with definite tumor masses, in all of whom biopsies and P³² tests were positive, 2) five patients with recognizable submucosal nodules, one of whom was not biopsied, the other four showing three positive, and one negative histologic diagnosis, all of whom had positive P³² tests, and 3) 13 patients in whom tumor could not be identified grossly, three of whom had positive, and seven negative biopsies, with three not biopsied. The two false negative P³² tests occurred in this group, the remaining 11 having definitely increased radiation consistent with a positive reading although the tumor in each case was submucosal extension from a gastric primary (Table 1).

Esophagitis. There were two false positive results in 22 patients, for an accuracy of 90.1 per cent. The two patients giving positive readings had severe, long-standing lesions with marked inflammatory thickening of the esophagus on resection.

Achalasia or cardiospasm. Tests in eight patients yielded one false positive, for an accuracy of 87 per cent. Thoracotomy and biopsy were carried out in the patient with the false positive reading, and there was no doubt of the error.

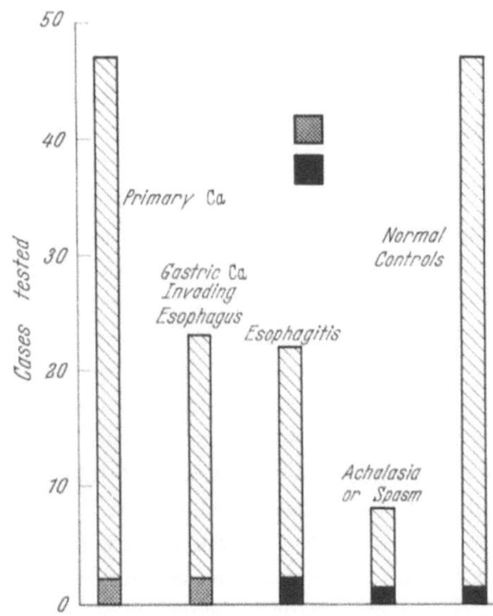

Fig. 3. The P³² test demonstrates a high degree of accuracy in the esophagus, as shown by the graph. Few errors are noted in any category of lesion. Results of 146 esophageal tests. Cancer accuracy 95.7%; overall accuracy 94.5%. O False negative, ● False positive

Normal controls. In 46 patients selected as normal controls, there was one false positive test, for an accuracy of 97.8 per cent.

Overall accuracy in the whole group of 146 patients was 94.5 per cent (Fig. 3).

Case Reports

Case 1. A 61 year old white male had complained of difficulty in swallowing foods for a period of ten months. This was associated with a 30 pound weight loss. A roentgenogram showed an irregular constricting lesion of the esophagus, present in the midesophagus, measuring 8 cm. in length, diagnosed as carcinoma. At esophagoscopic examination, a large cuff of neoplastic tissue arising from the right lateral wall of the esophagus was noted at 30 cm. The scope could be passed for 4 cm. beyond this point where a high degree of obstruction was reached. P³² readings were taken which dropped off sharply at 30 cm. A biopsy showed squamous cell carcinoma. The patient was treated with 5-Fluorouracil and x-ray grid therapy because of invasion of the bronchus noted on bronchoscopy. During irradiation therapy, he developed symptoms of bronchopleural fistula, and a cervical esophagstomy and gastrostomy were carried out. The patient did fairly well following this procedure, and gained ten pounds in weight. He is still on follow-up (Fig. 4).

Case 2. A 68 year old white female had noted dysphagia for approximately six months. X-ray examination showed carcinoma of the esophagus with shelving. At esophagoscopic examination large tumor masses were noted to block the lumen of the esophagus. The P^{32} test was markedly positive, and a biopsy was taken which showed squamous carcinoma grade III. Because of extreme emaciation, it was felt this patient was not in condition for surgery, and a feeding gastrostomy was performed. X-ray therapy was also felt not to be feasible. She died one year later of carcinoma of the esophagus (Fig. 5).

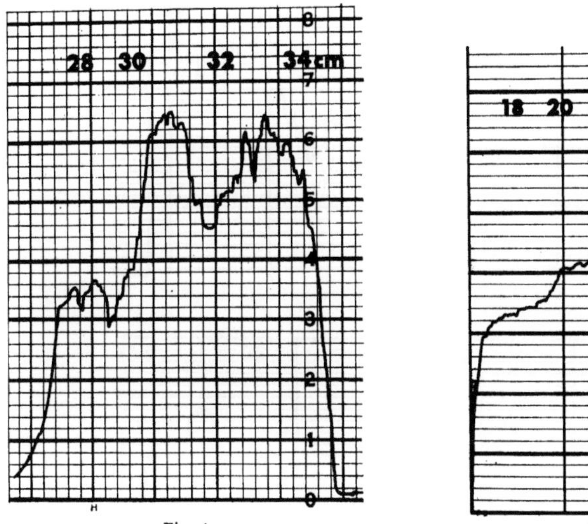

<table>
<tr><td>Fig. 4</td><td>Fig. 5</td></tr>
</table>

Figs. 4 and 5 (Cases 1 and 2). A marked increase in retained radiation is noted over the grossly and histologically cancerous lesions in these two patients

Case 3. A 60 year old white female, who had been treated three years previously for a squamous carcinoma of the cervix by x-ray therapy without recurrence, was seen complaining of dysphagia for three weeks. An esophagram showed a constricting lesion in the distal third of the esophagus with intrinsic involvement suggested. At esophagoscopic examination an infiltrative lesion was noted at 36 cm. with some narrowing, and P^{32} readings were strongly positive. A biopsy showed squamous carcinoma. Two years after resection she had no evidence of recurrence of carcinoma of the esophagus or cervix, and was doing well (Fig. 6).

Case 4. A 90 year old colored female had had dysphagia for two months. X-ray examination of the esophagus, three weeks previously, had shown an esophageal lesion. On esophagoscopic examination large masses of tumor were seen at a depth of 27 cm. P^{32} readings were taken, and the scope withdrawn after biopsy, which showed squamous cell carcinoma grade II. Because of her extreme age and general weakness no therapy was carried out. The patient died four months later without ever having been completely obstructed (Fig. 7).

Case 5. A 67 year old Spanish American female was seen complaining of six months' difficulty in swallowing and a weight loss of 25 pounds. She had been unable to swallow anything but liquids for the past month. X-ray examination had shown an annular lesion of the esophagus. At esophagoscopic examination, large exophytic masses were noted arising from the esophageal walls, but the tip of the scope could be passed beyond the lesion, and readings were taken distal to, through the tumor area, and proximally. Definite positive readings were noted over the tumor area. A biopsy was taken which showed squamous carcinoma grade II. The tumor was resected at thoracotomy, and an esophagogastrostomy and splenectomy were performed. She tolerated the procedure well, and was in excellent clinical condition three months later on follow-up (Fig. 8).

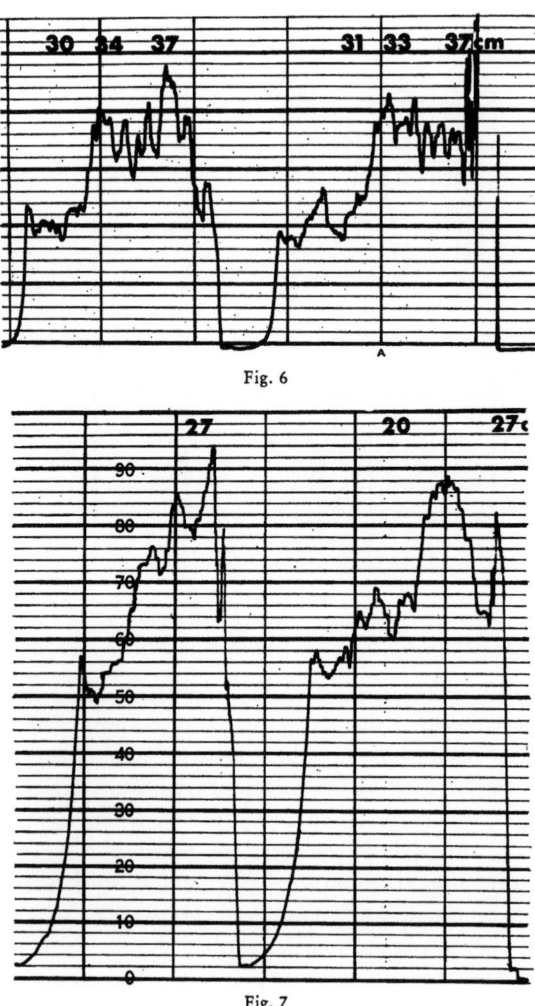

Fig. 6

Fig. 7

Figs. 6 and 7 (Cases 3 and 4). The marked radiation differential recorded in exophytic cancer of the esophagus may be duplicated at the same examination, as carried out in these two patients

Case 6. A 48 year old white male had difficulty in swallowing for four months with a sensation of obstruction beneath the xiphoid. He could swallow liquids but no solids. An x-ray examination showed circumscribed lesion at the junction of the middle and lower third of the esophagus designated as carcinoma. At esophagoscopic examination linear nodular protrusions of the esophageal wall were noted with the appearance of carcinoma. P³² readings were strongly positive. Biopsies showed marked chronic inflammation and atypical squamous carcinoma. Thoracotomy and esophagogastrostomy were carried out essentially uneventfully. Ten months later he was still swallowing fairly well, but complaining of pain in the left costal area. One month later he died suddenly of a hemorrhage from the esophagus (Fig. 9).

Case 7. This 66 year old white male had had a resection of the tongue for squamous cell carcinoma with postoperative irradiation, and had done quite well.

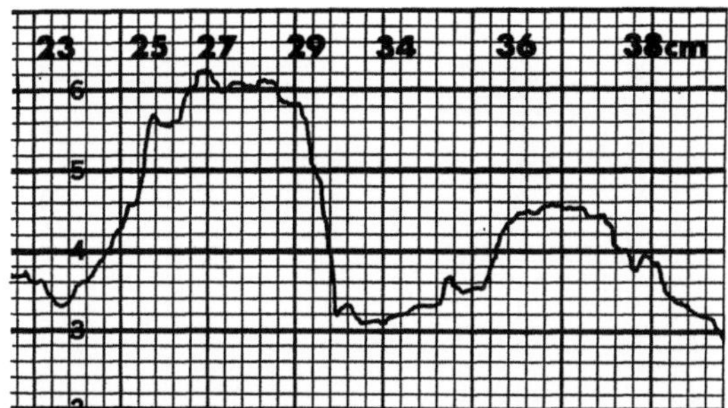

Fig. 8 (Case 5). When the esophagoscope may be passed beyond the lesion, it is possible to delineate the extent of the cancerous area rather precisely (from 25 to 29 cm as measured from the incisor teeth, in this instance)

Five years after his original surgery, he reported difficulty in swallowing with progressive dysphagia. An esophagram showed a lesion of the middle third of the esophagus. The first esophagoscopic examination, with biopsies, showed only chronic inflammation of the esophageal submucosa. Four days later, a second eso-

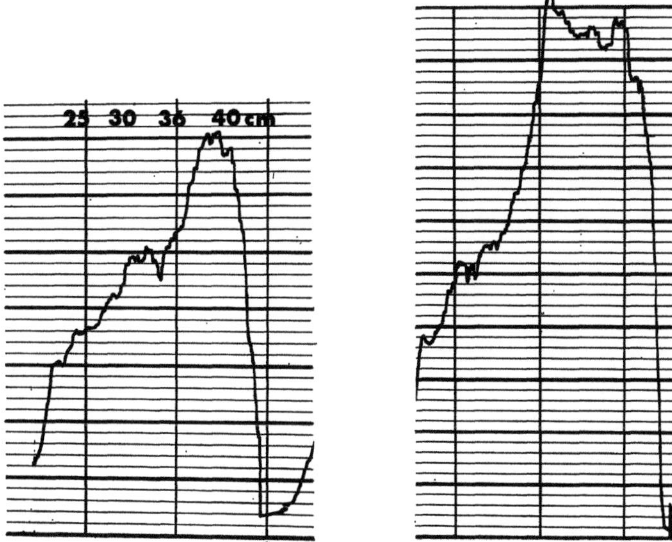

Fig. 9   Fig. 10

Figs. 9 and 10 (Cases 6 and 7). If the initial biopsy is negative, the positive P^{32} test provides additional evidence to support re-biopsy or surgical intervention as demonstrated in these two cases

phagoscopic examination was carried out, which revealed tumor masses occluding the lumen of the esophagus. P^{32} readings were taken which were highly positive, and a biopsy at this time showed squamous carcinoma grade III. The patient's tumor was resected and esophagogastrostomy carried out. Eighteen months later he felt perfectly well and was swallowing with ease (Fig. 10).

Case 8. A 62 year old white male had noted difficulty in swallowing for a period of five months, and had lost 14 pounds in weight. An esophagram showed a narrowing of the esophagus diagnosed as carcinoma. At esophagoscopic examination a polypoid tumor of the esophagus could be seen which constricted the lumen to some

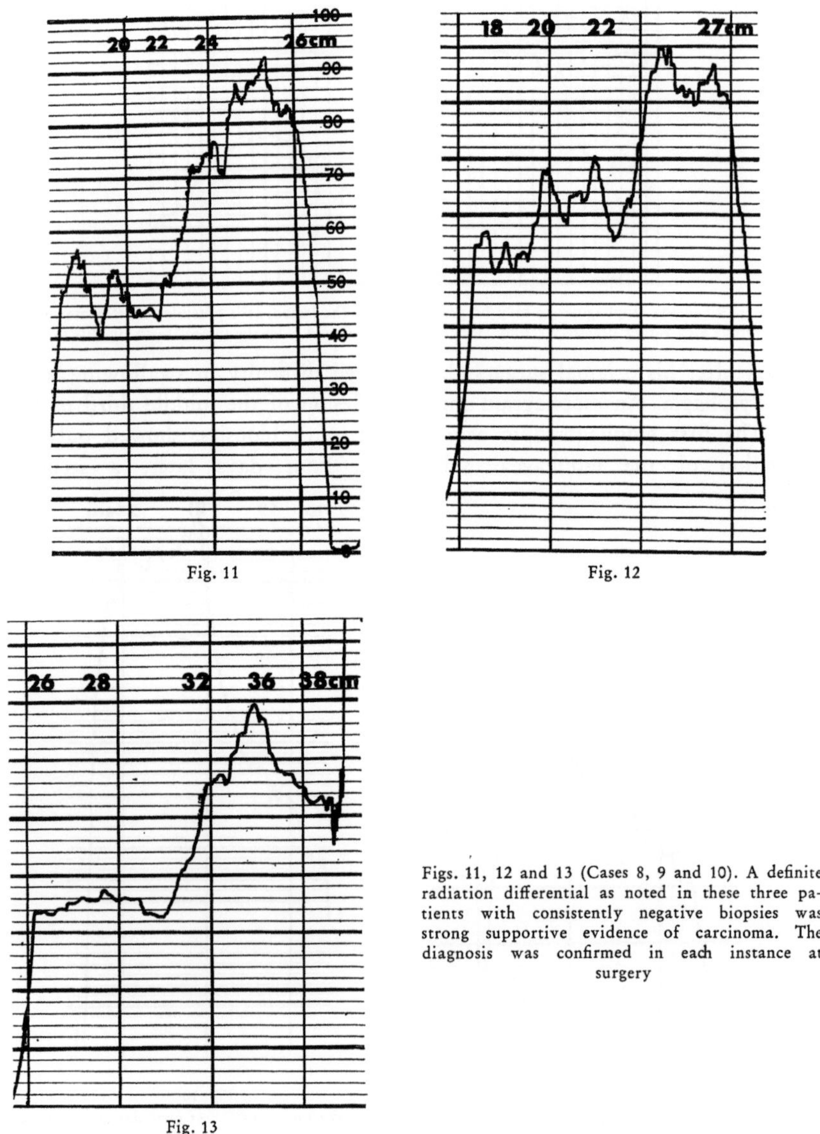

Fig. 11

Fig. 12

Fig. 13

Figs. 11, 12 and 13 (Cases 8, 9 and 10). A definite radiation differential as noted in these three patients with consistently negative biopsies was strong supportive evidence of carcinoma. The diagnosis was confirmed in each instance at surgery

extent, but the P³² counter could be passed beyond the point of tumor, and tests were run which were strongly positive. Biopsies showed only chronic esophagitis with marked atypical epithelial hyperplasia. The patient underwent thoracotomy and resection of the lesion which was diagnosed histologically as squamous cell carcinoma, but suffered cardiac arrest and died suddenly (Fig. 11).

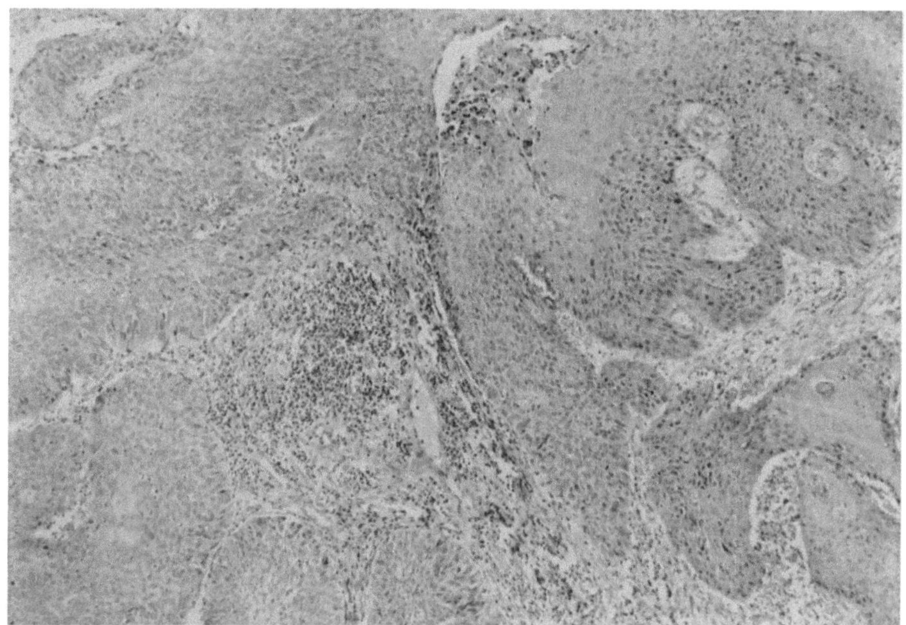

Fig. 14

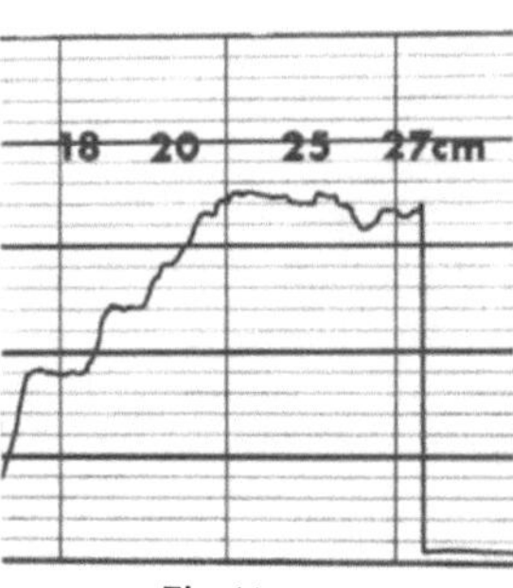

Fig. 16

Fig. 14 (Case 10). Squamous carcinoma resected at operation

Fig. 15 (Case 10). X-ray appearance of esophagus just prior to surgery. No definite tumor mass could be seen

Fig. 16 (Case 11). Three and one-half years following successful resection of carcinoma, a positive P^{32} test indicated recurrent cancer, although the gross appearance was that of esophagitis. A biopsy taken in the area of highest radiation revealed squamous cell carcinoma

Fig. 15

Case 9. A 61 year old white female had noted dysphagia with a feeling of obstruction in the lower sternal area just above the xiphoid for approximately 11 months, plus a 30 pound weight loss. She was able to take liquids only at the time of first being seen. On esophagoscopic examination, tumor was noted at 24 cm. and P³² readings were definitely positive. A biopsy was negative. X-ray examination showed cancer of the esophagus. At thoracotomy, the tumor was inoperable because of adherence to the right main stem bronchus and vessels in the right hilum. A feeding gastrostomy was placed in the stomach and systemic therapy with 5-FUDR was started. She did poorly, developed a tracheo-esophageal fistula, and was transferred home for terminal care (Fig. 12).

Case 10. A 75 year old Spanish American female had suddenly developed dysphagia one month prior to being seen. She had difficulty in swallowing both liquids and solids, and began to lose weight. Roentgenograms showed a constriction in the distal esophagus just above the cardia with partial obstruction suggesting the presence of carcinoma, but a definite diagnosis could not be made. At esophagoscopy, although no definite tumor could be made out, the lumen was occluded. However, the P³² counter could be passed to a depth of 39 cm. and markedly positive readings were obtained for approximately three to four centimeters. Biopsies were negative. Because of this fact, a second esophagoscopy was done one week later, and two more biopsies obtained which were also negative. At thoracotomy a squamous cell carcinoma of the esophagus, margins free of tumor, was resected. Lymph nodes were all negative, and a partial esophagectomy, proximal gastrectomy and esophagogastrostomy carried out. One year later, she was doing well and gaining weight (Figs. 13, 14, 15).

Case 11. A 65 year old white female had noted difficulty in swallowing, weakness, loss of weight and appetite over a period of one year, which had been gradually progressive. X-ray examination showed a lesion in the midesophagus with destruction of mucosa. At esophagoscopic examination a large polypoid tumor could be seen, and a biopsy showed squamous cell carcinoma. Esophagectomy and esophagogastrostomy were carried out. Following recovery, the patient did quite well although she developed pernicious anemia and was treated for this. Three and one-half years later, she was admitted with symptoms of regurgitation and poor food intake although no difficulty in swallowing. At esophagoscopy the esophagus appeared to be edematous and very friable with considerable bleeding. P³² readings were taken which were definitely positive, although the gross appearance of the esophagus was felt to represent esophagitis. An upper GI series was essentially within normal limits. A biopsy showed squamous cell carcinoma. No further treatment was given and she died two months later of her disease, almost four years after her initial surgery (Fig. 16).

Case 12. A 57 year old white male had experienced intermittent chest pains in the substernal area for approximately two years. Dysphagia and weight loss began seven months prior to being seen. Roentgenograms had shown a constrictive lesion involving the thoracic esophagus. On esophagoscopic examination, three smooth masses of tumor were noted projecting from the walls of the esophagus at 28 cm. The P³² test was markedly positive, and the biopsy showed carcinoma of the esophagus. A palliative esophagectomy, esophagogastrostomy and pyloroplasty were carried out, but some tumor was left behind. The patient did fairly well although be continued to complain of chest pain, but was swallowing well. Three months after the initial study, a second esophagoscopy was carried out, and necrotic bleeding tissue could be seen at 22 cm. A positive P^{32} test was noted at this point, but dropped to background levels at 16 cm. X-ray therapy was scheduled but the patient suddenly became worse, deteriorated rapidly and died (Figs. 17 and 18).

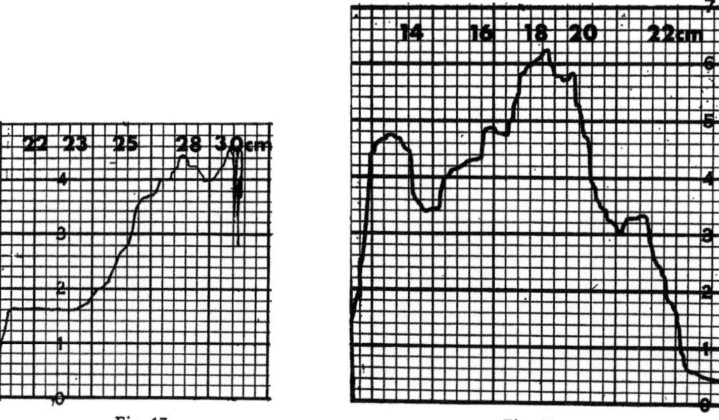

Fig. 17 Fig. 18

Figs. 17 and 18 (Case 12). Three months after a palliative resection for cancer, preceeded by a positive P^{32} test (Fig. 17), a second test demonstrates residual tumor tissue (Fig. 18)

Case 13. A 66 year old white male had noted dysphagia for approximately three months. This was largely with solid foods, and he had had no difficulty with liquids. An initial roentgenogram showed an intraluminal polypoid lesion of the distal esophagus extending to the fundus of the stomach, approximately 7 cm. in length, diagnosed as carcinoma. At esophagoscopy, the instrument was passed with ease to a depth of 40 cm. where a large tumor mass was seen projecting from the right lateral wall of the esophagus partly filling the lumen. P^{32} readings were taken which were markedly positive. A biopsy showed poorly differentiated squamous cell carcinoma. An esophagectomy and proximal gastrectomy were performed with an esophagogastrostomy and primary anastomosis below the aortic arch. The lesion was grossly involving the proximal one-third of the stomach. Regional lymph nodes were negative for tumor. Following surgery, the patient did fairly well for several months then began to complain of nausea and vomiting after meals as well as excessive salivation. Five months after his original surgery, an esophagoscopy performed by the Thoracic Surgery Service was within normal limits. Roentgenograms taken at this time also showed no evidence of recurrent disease in the esophagus. However, five days later, a repeat esophagoscopy and P^{32} test gave marked elevation of radiation in the region of the anastomosis at 30 cm. Grossly, the area looked equivocal for recurrence. A biopsy was not taken. The patient died six weeks later of his disease (Figs. 19 and 20).

Case 14. A 48 year old white male had been in good health until three months previously when he noted a slowly progressive dysphagia. X-ray examination showed thickening and narrowing of the esophagus approximately 4 cm. above the cardia. The official reading on the outside films was esophagitis with ulceration; no definite tumor mass identified. A new x-ray study was obtained which was again felt to show esophagitis with ulceration but no tumor, and on departmental review, all but one staff member concurred in this opinion (Fig. 21). On esophagoscopic examination, the esophagoscope was passed into the stomach with ease and no definite tumor

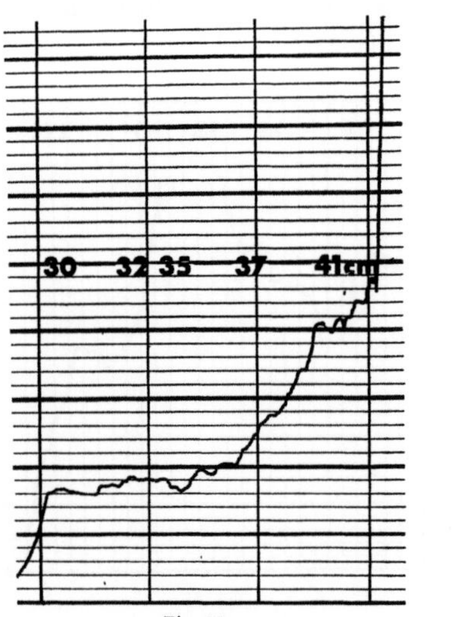

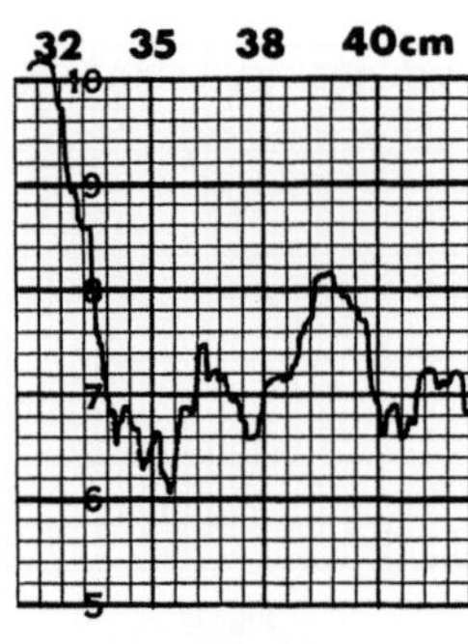

Fig. 19Fig. 20

Figs. 19 and 20 (Case 13). The marked differential in radiation noted prior to surgery (Fig. 19) was again noted five months later on follow-up, despite lack of roentgenographic and esophagoscopic evidence for recurrence (Fig. 20). Death from carcinoma of the esophagus six weeks later indicated an accurate result

formation could be made out, although there was a slight sense of resistance at the cardioesophageal junction. A P³² test was positive (Fig. 22) and biopsies were taken at 40 and 38 cm., which histologically showed unclassified adenocarcinoma in submucosa (Fig. 23). On thoracotomy, a tumor mass was noted in the distal portion of the esophagus adjacent to the stomach. This was resected and esophagogastrectomy, gastrostomy and pyloromyotomy were carried. The pathological specimen, which showed the terminal esophagus and cardioesophageal junction, revealed adenocarcinoma involving these tissues with metastases to three regional lymph nodes (Fig. 24). The patient initially did well following surgery, but five months later complained of some difficulty in swallowing again and dilations were carried out, which were only partly successful in relieving these symptoms. X-ray finding at this time showed no recurrence of tumor (Fig. 25). Six months after the initial resection, another esophagoscopy showed considerable inflammatory change at the gastroesophageal anastomotic junction. P³² test was definitely elevated, but a biopsy showed only inflammation. Following this procedure, he seemed to improve considerably over the next few months, and it was felt that the only positive evidence for recurrence was the P³² test (Fig. 26). A repeat esophagoscopic examination four months later showed a definite elevation diagnostic of tumor over a distance of 2 cm. near the junction of the esophagus with the stomach (Fig. 27). The area appeared nodular and the diagnosis was probable recurrence at the anastomosis. Biopsy showed adenocarcinoma in submucosa (Fig. 28). After a short delay, the patient was given 5-Fluorouracil therapy by way of intra-aortic catheterization through the left femoral artery over a period of 21 days for a total dosage of 6.15 g. Very little, if any, response was noted to this type of treatment. The patient developed a right Horner's syndrome with a hard, fixed nodule beneath the lower end of the right sternocleido-mastoid. It was felt that radiotherapy at this point would be unavailing and a gastrostomy probably would be necessary.

Case 15. A 58 year old white male had had a total gastrectomy six years previously for cancer of the stomach. A very small cuff of gastric tissue had been left just below the cardioesophageal junction. Over a period of seven months, he had noted difficulty in swallowing. Food stuck in the lower esophagus, and on regurgitation a large plug of mucus came up with it. There had been a 12 pound weight loss. A recent esophagram was reported as showing no definite pathology. On esophagoscopic examination, a rough collection of tissue appeared on the right lateral wall of the esophagus at a depth of 34 cm. The eso-phagoscope could be passed beyond this area with-out any difficulty to a depth of 42 cm. The gross appearance of the area however was that of mucosal infiltration with tumor. P^{32} readings from 42 to 34 cm. were definitely positive. Three biopsies were taken which showed mucin-producing adenocarci-noma. The patient was returned to his physician for surgery (Figs. 29 and 30).

Case 16. A 59 year old colored male had noted postprandial pain for approximately $1^{1/2}$ years. A roentgenogram showed a lesion of the upper portion of the stomach with dilatation of the distal portion of the esophagus. There had been a 22 pound weight loss. The esophagoscope was passed to a depth of 43 cm. where large polypoid folds could be seen. P^{32} readings were taken which were positive, and biopsies were taken, which showed only esophageal tissue, no carcinoma present. At laparotomy, an adenocarcinoma of the stomach was found, extend-ing into the lower esophagus and penetrating the serosal surface. Lymph nodes were negative for tumor. A total radical gastrectomy, distal esophag-ectomy, splenectomy and esophagojejunostomy were performed. Eight months later, he had gained 10 pounds, and was asymptomatic (Fig. 31).

Case 17. A 52 year old white male was seen with a one month's difficulty in swallowing solid foods. There was pain on swallowing but the patient could usually get food down if he took water with it. In 1962, he had had a subtotal gastrectomy for bleeding peptic ulcer. A recent esophagram was read as showing probable car-cinoma in the distal third of the esophagus. On esophagoscopic examination, large masses of tissue were seen partly obstructing the lumen of the esophagus at 37 cm.

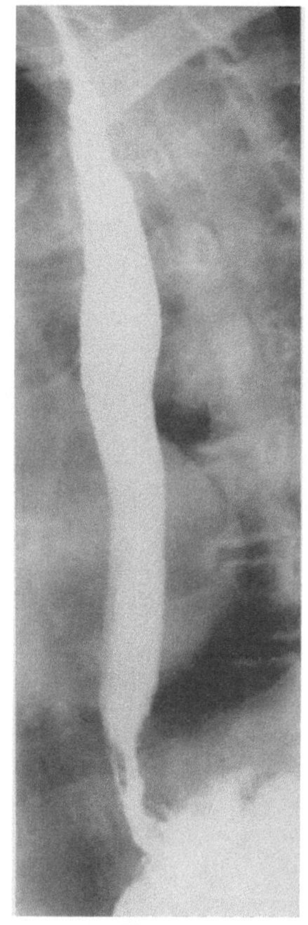

Fig. 21

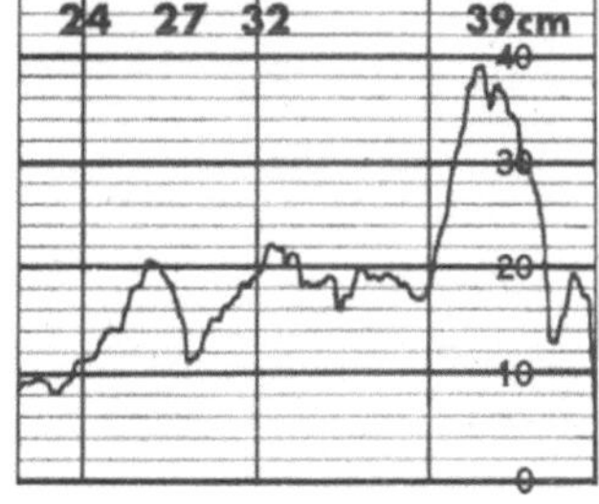

Fig. 22

The tip of the scope could be passed beyond this area to 40 cm. where a severe inflammatory reaction and moderate ulceration could be made out. P^{32} test showed markedly positive readings at the 36 cm. level. The gross diagnosis was carcinoma of

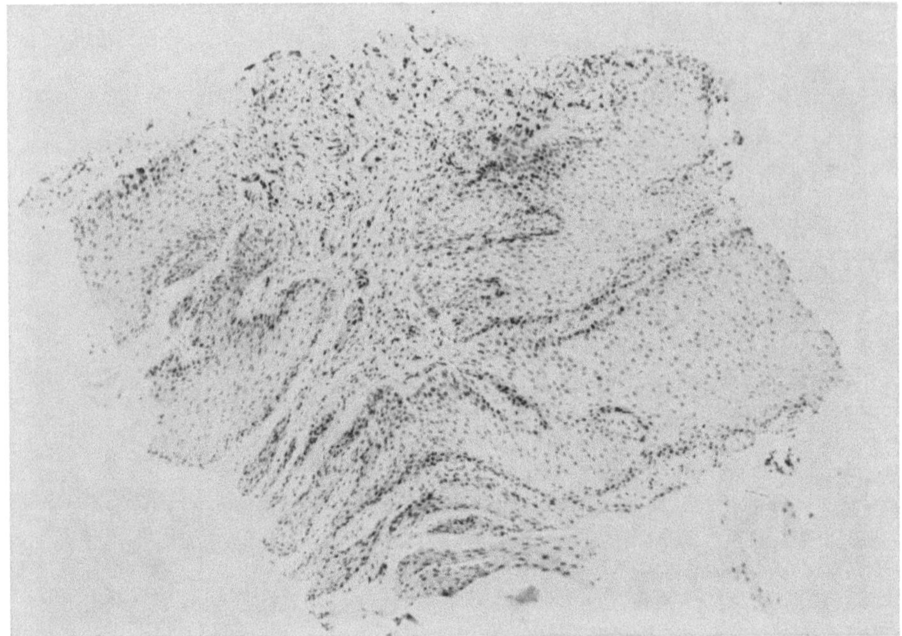

Fig. 23

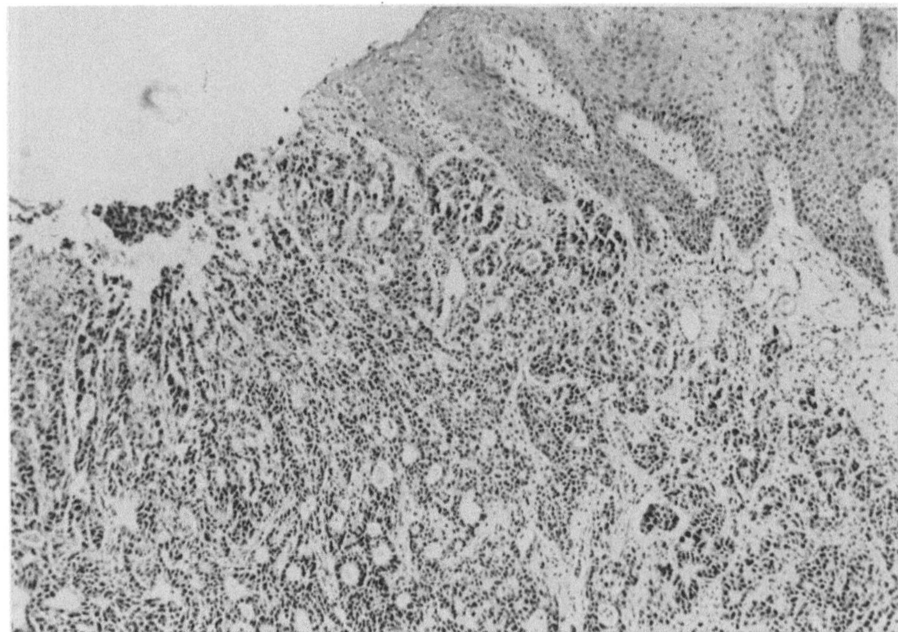

Fig. 24

Figs. 21, 22 and 23 (Case 14).Original roentgenogram which was interpreted as showing esophagitis with ulceration (Fig. 21), positive P^{32} test (Fig. 22), and biopsy (Fig. 23) obtained at esophagoscopy and resected specimen (Fig. 24)

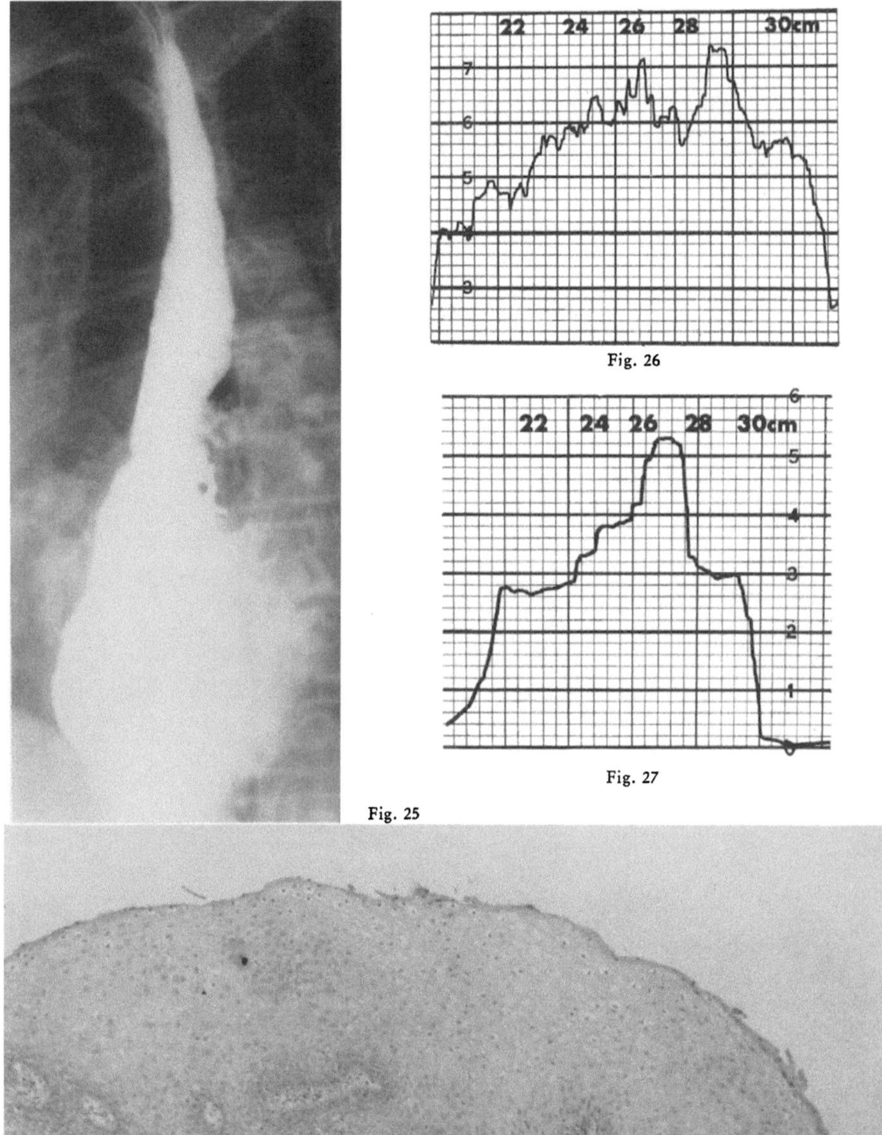

Fig. 26

Fig. 27

Fig. 25

Fig. 28

Figs. 25 and 26 (Case 14). Esophagram obtained five months after surgery (Fig. 25) and positive P^{32} test one month later (Fig. 26)

Figs. 27 and 28 (Case 14). Persistently positive P^{32} test (Fig. 27) and positive biopsy (Fig. 28) 10 months after resection, despite lack of roentgen evidence of recurrence

the esophagus. Biopsy showed poorly differentiated adenocarcinoma. At thoracotomy, a carcinoma of the cardia of the stomach with extension into the esophagus and to the celiac and aortic nodes were found. Esophagogastrostomy and esophago-

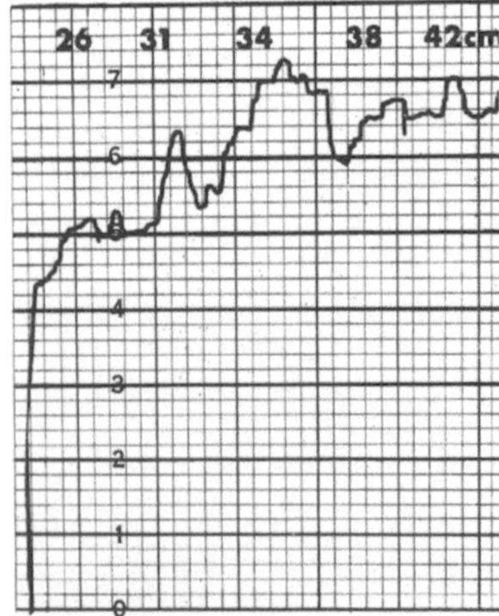

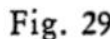

Figs. 29 and 30 (Case 15). Six years after total gastrectomy for adenocarcinoma, a positive P³² test and biopsy were demonstrated at esophagoscopy in a patient with negative findings on roentgenoscopy

Fig. 29

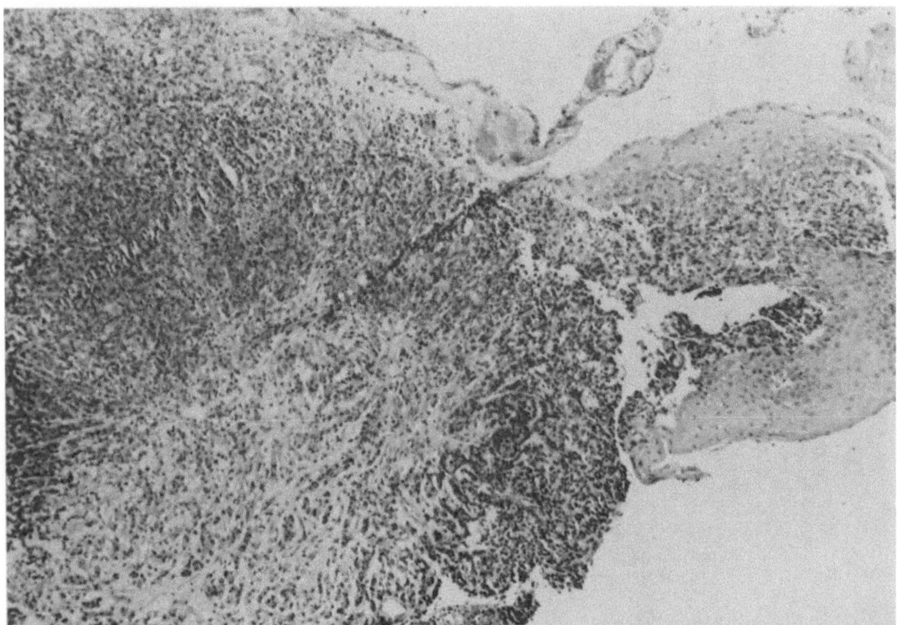

Fig. 30

gastrectomy were carried out. The patient was released to his referring physician for x-ray therapy (Fig. 32).

Case 18. A 52 year old white male had noted increased difficulty in swallowing over a period of four months. On admission, he was unable to swallow fluids without considerable difficulty. On esophagoscopic examination, nodular linear lesions were noted at a depth of 35 cm. on the anterior and right lateral wall of the esophagus. The esophagoscope could be passed beyond this area to a depth of 40 cm. The P^{32} readings were positive within this area. Biopsy showed mucin-producing adenocarcinoma in squamous submucosa. On x-ray examination, there was marked

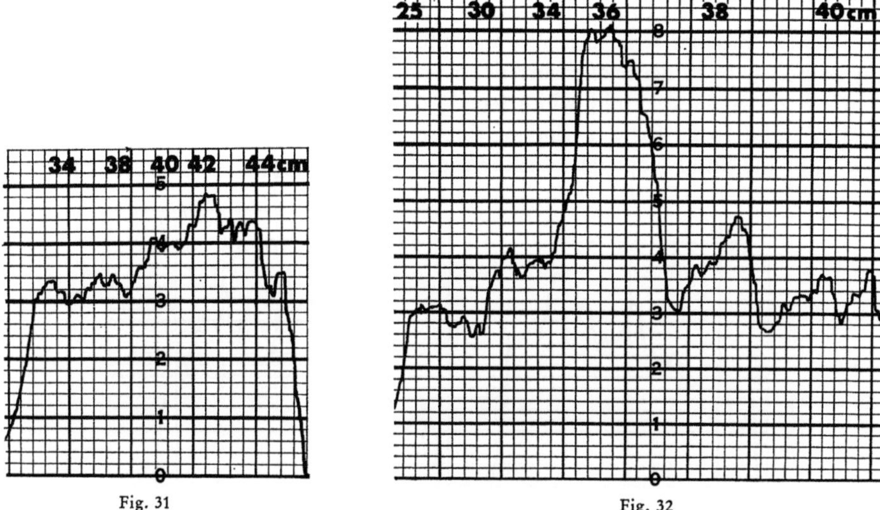

Fig. 31   Fig. 32

Fig. 31 (Case 16). A positive radiation differential was obtained over an area of submucosal invasion from carcinoma of the stomach. The biopsy in this case was negative, as were the gross findings, and the diagnosis proven only at laparotomy

Fig. 32 (Case 17). Although this adenocarcinoma of the stomach infiltrated the submucosa of the esophagus for a considerable distance, the only true evidence of mucosal tumor was seen at 36 cm, as noted also on the P^{32} test

irregularity of the fundus and body of the stomach, although no pathology was noted in the esophagus. The findings were grossly interpreted as being more likely due to lymphoma than carcinoma. At the time of initial examination, abdominal fluid was noted and on paracentesis, fluid was removed which, on cell block, showed malignant cells, class V. The patient was placed on 5-Fluorouracil therapy and chromic phosphate, P^{32}, was instilled to help control the ascites. In all, the patient received four courses of 5-FU, but eventually began to deteriorate and expired nine months after the initial diagnosis (Fig. 33).

Case 19. This 59 year old Spanish American male had noted dysphagia increasingly over a period of six months. Three months prior to admission, an esophagram had failed to reveal pathology. However, more recent films showed a constricting lesion, probably carcinoma of the distal esophagus and cardiac portion of the stomach. On esophagoscopic examination, a suspicious area was found at the lower end of the esophagus. P^{32} reading was positive, but the biopsy showed only chronic esophagitis with epithelial atypism. Gastric washing, however, showed class V cells, indicating malignancy and a new x-ray examination revealed a constricting lesion, probably carcinoma, of the distal esophagus and cardiac portion of the stomach. At laparotomy, a tumor mass of the distal esophagus and cardiac area of the stomach was found without apparent metastases. A radical proximal gastrectomy, splenectomy, pyloroplasty, distal esophagectomy, and esophagogastrostomy were performed. The patient made an uneventful recovery from this procedure. However, on follow-up one year later, there was x-ray evidence of possible recurrence in the gastric pouch and on gastroscopic examination, as well as esophagoscopic examination, a heavy ridge of tissue was noted in the pouch which was felt to be probable recurrent carcinoma. An attempt was made to perform a P^{32} test on this occasion, but it was felt to be unsatisfactory. At re-exploration, no tumor was found in the stomach. Biopsies were negative, and subsequent examinations have failed to show any pathology or evidence of recurrence over a period of three years (Fig. 34).

Case 20. A 68 year old white female complained of vomiting and inability to swallow solid foods for a period of three weeks. A roentgenogram showed a narrowing of the distal esophagus with irregularity of the upper fundus suggesting car-

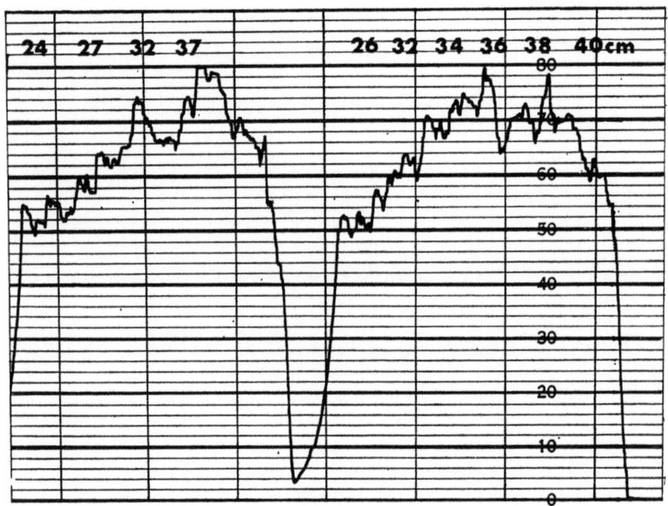

Fig. 33

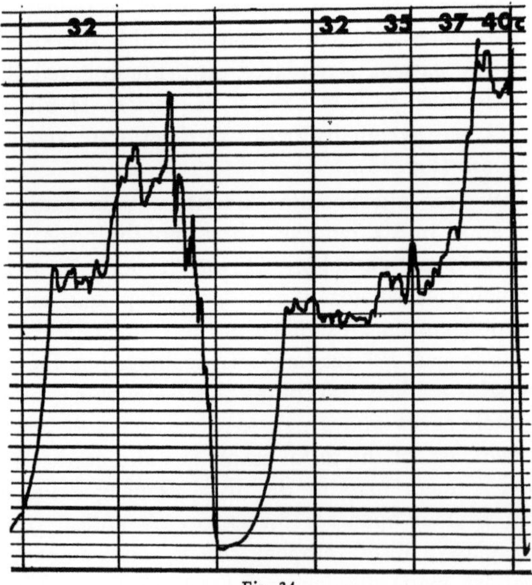

Fig. 34

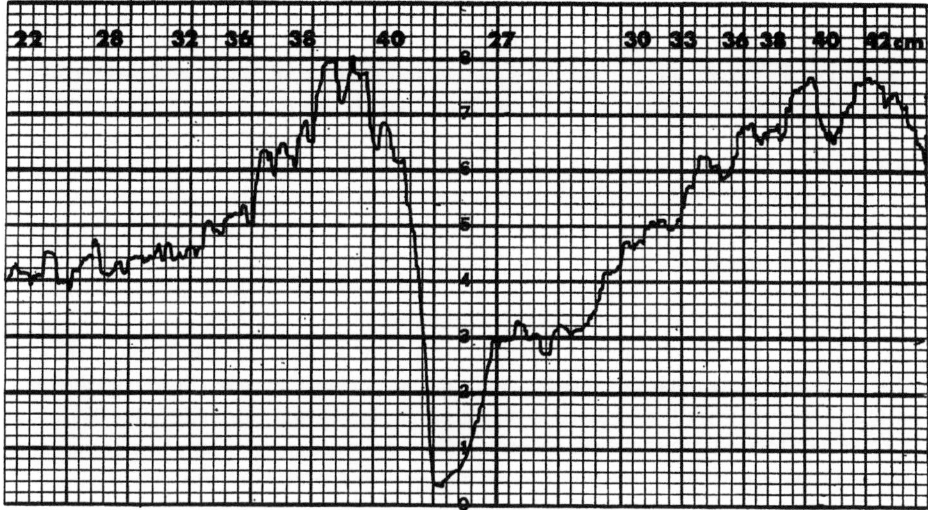

Figs. 33, 34 and 35 (Cases 18, 19 and 20). Reproducible radiation differentials were noted in three patients with adenocarcinoma of the stomach infiltrating the esophagus submucosally. Biopsy was positive in one (Fig. 33), and negative in two others (Figs. 34 and 35)

cinoma of the fundus extending into the lower esophagus. At esophagoscopy, the instrument could be passed with ease to a depth of 41 cm., and no definite tumor was identified. A P^{32} test was taken in duplicate, and both readings showed a markedly elevated count from approximately 42 cm. to 32 cm. Biopsies taken showed esophageal tissue with microscopic islands of distorted atypical cells in submucosa, probably representing tumor cells. At exploratory laparotomy, a massively invading carcinoma of the esophagus and stomach was found, which was determined to be nonresectable. A Mousseau-Barbin esophagogastric feeding tube was inserted, however, the patient did poorly, and expired one month later of her disease (Fig. 35).

Case 21. A 63 year old white male had been treated for leukoplakia of the tongue and floor of the mouth. He also had complaints of pain postprandially over a period of several years. An esophagram showed a medium size hiatal hernia but no other abnormal findings, and the stomach and duodenum were essentially negative. He was treated symptomatically with antacids and did fairly well over a period of six years when he suddenly began to have much more severe pain, and considerable dysphagia. Evidence of erosion of the esophagus was noted at 36 cm. P^{32} study was negative. The region was dilated, and a biopsy showed only chronic inflammation. Despite continued medical therapy, he

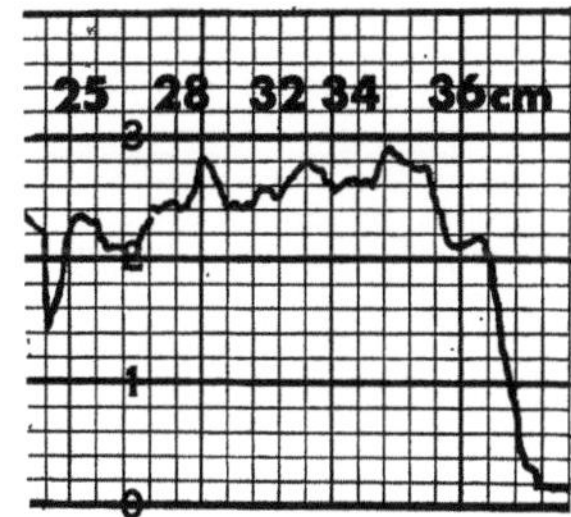

Fig. 36 (Case 21). Background radiation noted in a patient with long-standing esophagitis with stricture, which eventually required surgical intervention

continued to have a high degree of obstruction, and eventually underwent thoracotomy with esophagectomy, an esophagogastrostomy and pyloroplasty with the finding of an acute ulceration at the esophagogastric junction, benign. Following this procedure, he did only fairly well, developed hepatitis, but eventually recovered completely and was sent home in good condition for follow-up (Fig. 36).

Case 22. A 70 year old white female had noted dysphagia of one year's duration which had progressed within the past three months to the point where she was unable to swallow any solid food. X-ray examination showed a narrowing of the distal third of the esophagus just above a hiatus hernia with a suggestion of ulcer in the area of narrowing. The diagnosis was probable esophagitis with possible ulceration in the esophagus. It was felt to be most likely on an inflammatory basis. On esophagoscopic examination, there appeared to be a large submucosal mass pushing the mucosa in front of the esophagoscope at 35 cm. P³² readings were definitely positive over the area, and a biopsy was taken which was nondiagnostic showing only esophageal mucosa. At thoracotomy, carcinoma was not found, but there was marked fibrosis, chronic inflammation and hypertrophy of smooth muscle. Postoperatively, she did exceedingly well, and at time of discharge was eating normally (Fig. 37).

Case 23. This 61 year old white male was seen with intermittent symptoms of blockage, over a period of two months, at the level of the xiphoid process when trying to swallow solid food. When seen, he could swallow liquids only, and had lost 12 pounds in the past two weeks. A roentgenogram showed a local area of narrowing in the lower esophagus which had the appearance of a very early intramural lesion. The lung fields were within normal limits. At the initial esophagoscopy, the esophagoscope could be passed easily into the stomach and no abnormalities were noted. Following this procedure, the patient was able to swallow normally for the following week. The examination was repeated with a P³² test one week later, and again the esophagus was grossly negative, and the P³² test was also within normal

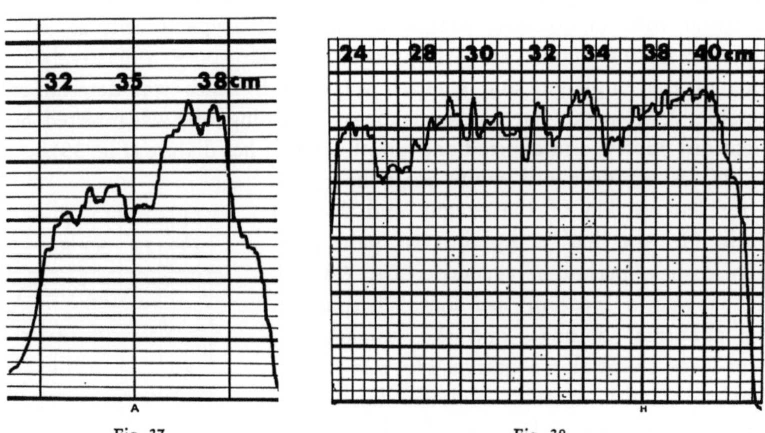

A
Fig. 37
H
Fig. 38

Fig. 37 (Case 22). A false positive P³² result was obtained in a case of chronic esophagitis with marked inflammatory reaction. No tumor was found when the inflammatory mass was resected

Fig. 38 (Case 23). This negative test, demonstrating only background radiation, was obtained in a patient with dysphagia and carcinoma of the lung. The tumor impinged upon, but did not invade, the lower esophagus

limits. Over the next two months, his symptoms increased again, and an esophagram showed more narrowing. A repeat esophagoscopic examination performed five months after the initial examination revealed an area of narrowing at 38 cm., but no mucosal invasion. Because of the equivocal findings, and yet the definite clinical symptoms of the patient, a thoracotomy was performed revealing carcinoma in the left lower lobe of the lung compressing the esophagus with implants to the pericardium, diaphragm, and visceropleura. A biopsy was positive for squamous cell carcinoma of the left lung. He was placed on x-ray therapy, but a gastrostomy was necessary four months later. He is still on follow-up (Fig. 38).

Case 24. A 54 year old white male was seen with a complaint of substernal discomfort of two months' duration. Symptoms consisted of a disagreeable sensation in the midsternal region on swallowing food which was accentuated by hot or cold materials. This had been previously evaluated by roentgenoscopy and esophagoscopy with negative findings (Fig. 39). However, on further examination an ulcerative lesion of the soft and hard palate of the right side was noted, and a biopsy had shown squamous cell carcinoma. X-ray therapy had been started to the palatal lesion, but because of continued complaints of substernal discomfort, esophagoscopy and P^{32} tests were carried out. On initial passage of the esophagoscope, no lesion was seen and there was no obstruction to the instrument to a depth of 42 cm. The P^{32} probe was then inserted and readings were taken from this level to approximately 20 cm. In the area of 30 to 25 cm. there were definite diagnostic readings (Fig. 40). The P^{32} probe was removed and the esophagus studied carefully in this area at which time a definite, small, exophytic tumor was visualized on the left lateral and posterior wall of the esophagus. Two biopsies were taken which showed squamous cell carcinoma (Fig. 41). X-ray therapy to the palate lesion was stopped. The patient

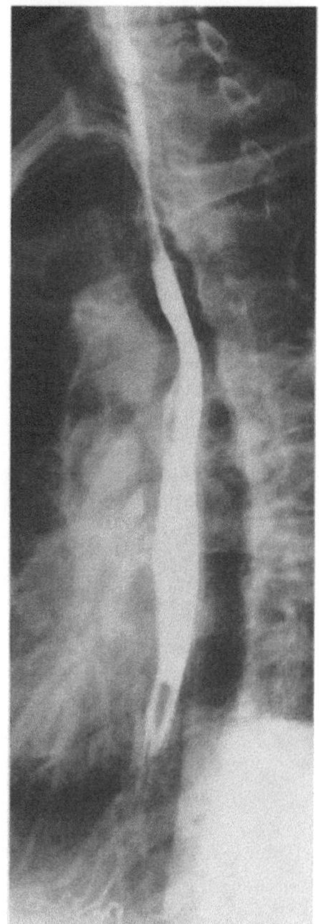

Fig. 39

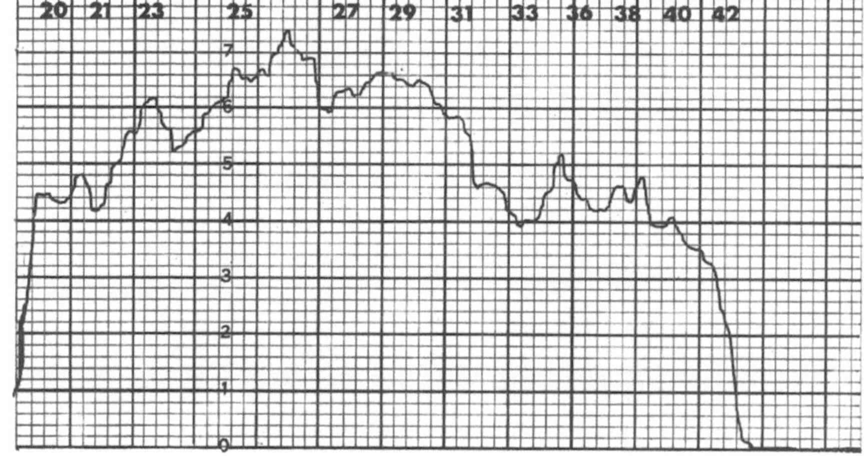

Fig. 40

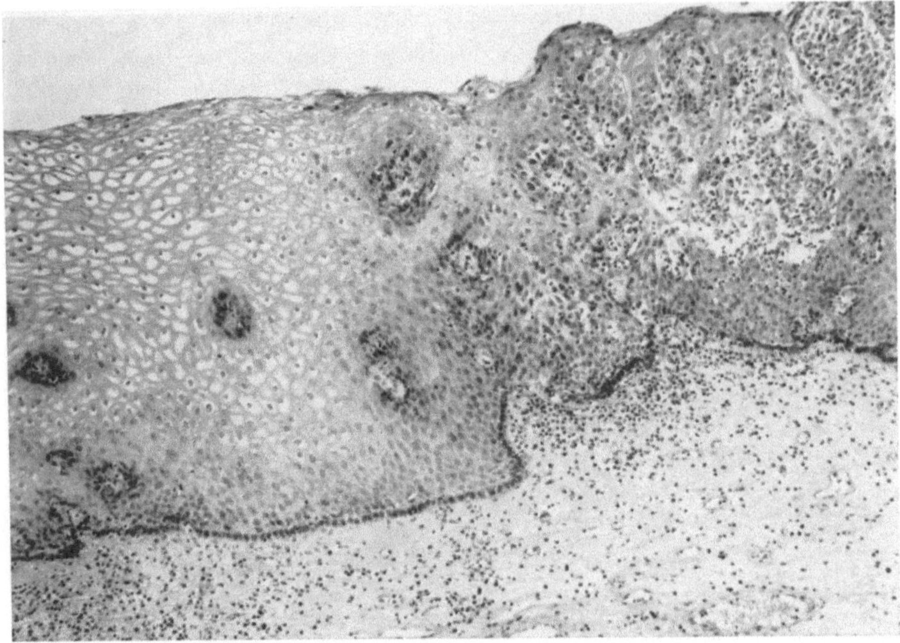

Fig. 41

Figs. 39, 40 and 41 (Case 24). Although the roentgenogram repeatedly showed no abnormalities (Fig. 39), the P^{32} test revealed the area of tumor (Fig. 40) which gave a positive biopsy (Fig. 41)

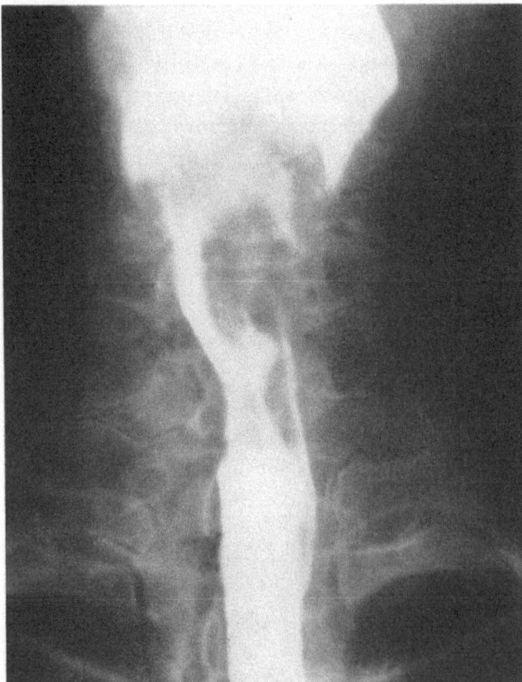

Figs. 42 and 43 (Case 25). Extrinsic pressure on the esophagus without gross mucosal involvement (Fig. 42) was diagnosed correctly as tumor by means of the P^{32} test (Fig. 43)

underwent esophagogastrostomy and pyloromyotomy for excision of a squamous carcinoma Grade III of the esophagus. The tumor infiltrated the superficial layer of the muscular coat but no tumor penetrated the muscularis. All lymph nodes were negative. Following surgery, the patient continued x-ray therapy to the palate lesion and after four months was doing exceedingly well.

Case 25. A 77 year old white male had noted a swelling in the left side of the neck of nine months' duration. A needle biopsy of the mass showed squamous cell carcinoma and fibrous tissue. An esophagram revealed a rounded rather sharpely defined soft tissue mass on the left lateral wall of the cervical esophagus in the region of C-7 and T-1 (Fig. 42). At esophagoscopy the tip of the instrument could be passed only to 18 cm. Further passage was prevented by massive deflection of the esophageal wall, apparently by tumor. There was no involvement of the mucosa of the eso-

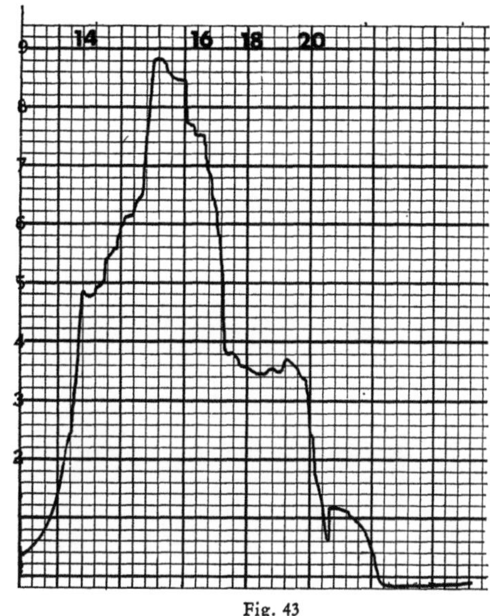

Fig. 43

phagus. The P^{32} counter could be passed to a depth of 20 cm. and on retraction of the instrument, there was a marked elevation at 16 cm. with background readings at 14 cm. above this level (Fig. 43). Although no biopsies were taken, the test appeared to prove conclusively the presence of carcinoma extrinsic to the esophagus without gross mucosal involvement. The patient was continued on x-ray therapy to the tumor mass with good results.

c) Stomach

Methods. The patient is prepared as for esophagoscopy. Two types of apparatus have been used. The original instrument, and the one employed in obtaining all results reported in the stomach, is the Eder gastroscope with Bernstein head and controllable tip. Manipulation of the tip of this instrument by a control at the handle allows approximation of the Geiger counter, inserted just distal to the lamp and objective window, to the area to be tested (Fig. 44). Since the counter is not exactly in line with the objective, but several centimeters distal to it, precise placing of the Geiger tube against the test area is often difficult. It may be impossible to be sure that an accurate result has been obtained in relatively small lesions, such as polyps and ulcers. Where there is a fairly large area of involvement, valid positive readings are obtained in the case of carcinoma, proving the feasibility of the P^{32} test for evaluation of the stomach. In small lesions, however, doubt may exist. Inflation of the stomach, so necessary for adequate visualization of most of the mucosal area, further complicates the situation by increasing the distances the tip of the gastroscope must travel to reach the desired area. It is also true that in a survey operation involving any large area of the gastric mucosa, difficulties are compounded by the inflation necessary to the procedure. In addition, antral lesions are almost invariably inaccessible to the tip of the gastroscope.

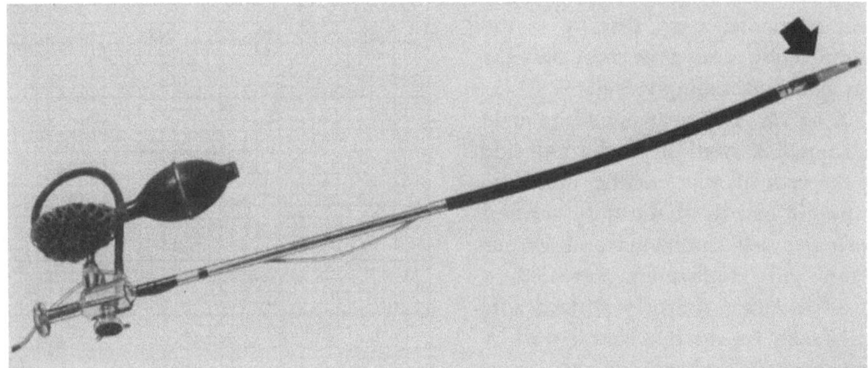

Fig. 44. The controllable-tip gastroscope is shown, the arrow designating location of the counter, distal to the lamp and objective window. The tip may be bent in two directions by manipulation of the small control at the proximal end

Because of these difficulties, an experimental apparatus was devised (Fig. 45) consisting of a short esophagoscope through which a thin gastroscope might be passed, with an additional groove for a separate controllable counter probe. The esophagoscope is passed first, held in place, and the gastroscope and probe next inserted

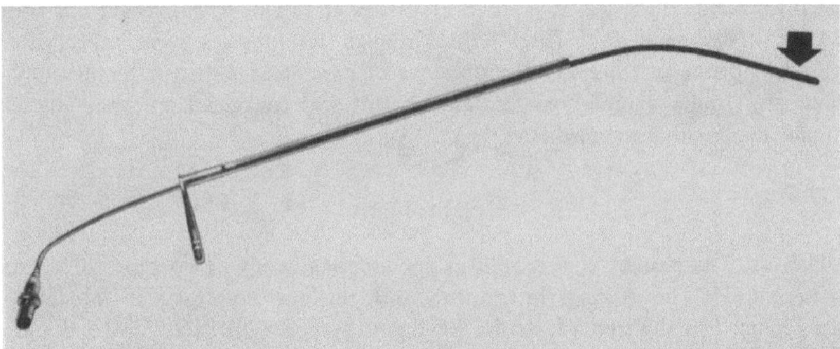

Fig. 45. The experimental apparatus shown consists of a short esophagoscope which carries inferiorly an adjustable counter probe. The arrow marks the position of the Geiger counter. The viewing gastroscope (not shown) fits in the esophagoscope above the counter probe. Only limited success was attained with this equipment

separately. The counter probe, which has a spiral metal sheath, is then manipulated under direct vision supplied by the gastroscope. This apparatus, while excellent in principle, failed to solve the problem of scanning the stomach in fairly extensive testing. The awkward situation involving the management of esophagoscope, gastroscope, and scanner at the same time proved quite difficult, but the real problem was the scanner probe, which was not sufficiently controllable in the stomach. No adequate tests were obtained with this apparatus, and trials with it are not given in results of stomach testing.

Results. In 56 tests in the same number of patients, four were unsatisfactory in that it was obviously impossible to adequately position the Geiger counter on the test area. These failures are mentioned only to point out the difficulty of carrying out the procedure objectively in the stomach. In all others, adequate approximation was obtained.

Cancer. There were five false negative tests in a group of 26, four of which were unsatisfactory, for an accuracy of 77.3 per cent. Most of these tumors involved a significant area of the gastric mucosa.

Benign ulcer. In 14 patients, there were four false positive findings, for an accuracy of 71.4 per cent.

Gastritis and benign tumors. One patient with severe atrophic gastritis, and a roentgenogram diagnostic of extensive carcinoma was correctly diagnosed as negative. Two with benign adenomatous polyps, in whom the Geiger counter was accurately approximated, failed to register increased radiation values.

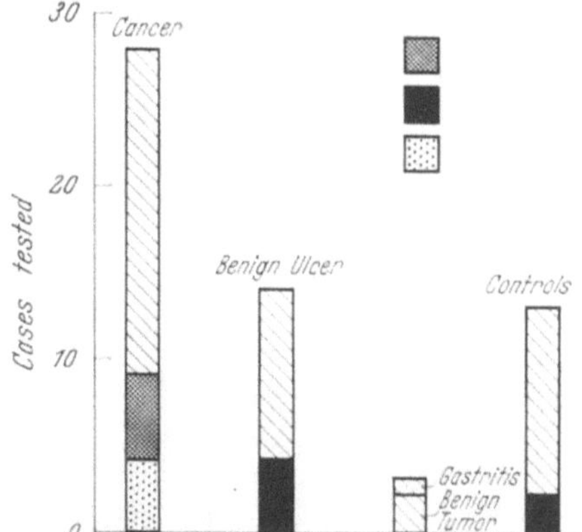

Fig. 46. Graph illustrates the rather poor results obtained with the P³² test in the stomach, largely due to technical difficulties in instrumentation. Results of 56 gastric tests. Cancer accuracy 77.3%; overall accuracy 78.8%. ○ False negative, ● False positive, ● Unsatisfactory

Normal controls. There were two false positive results in a group of 13 normal controls, for an accuracy of 84.6 per cent.

The overall accuracy of satisfactory gastric tests was 78.8 per cent (Fig. 46).

Case Reports

Case 26. The illness in this 54 year old white female began with intermittent vomiting which gradually became worse over a period of two months. Prior to admission here, a subtotal gastrectomy had been reportedly carried out, the pathological report on the resected tissue being malignant lymphoma, lymphosarcoma type. X-ray examination on admission showed probable antral obstruction, partial. There was no evidence of a resection having been made. At gastroscopic examination, the angulus and antrum could not be identified because of gross tumor deformity of the lower end of the stomach. The P³² test was positive. The patient was placed on radiotherapy, a dose of 4,150 rads being given over a five week period. There was no great change in her clinical condition, and she expired at home one month later (Fig. 47).

Case 27. This 52 year old white male had had anorexia and insidious weight loss, plus progressive weakness and fatigue six months prior to being seen. He had had no pain, and rare nausea or vomiting. X-rays made five months previously were diagnosed as peptic ulcer and treatment was instituted without significant improvement. Abdominal distention had appeared three weeks before being seen. A roentgenogram showed a large flat ulcer of the body of the stomach just above the incisura with stiffening of the stomach wall, compatible with carcinoma. At gastroscopic examination, it was impossible to make a good examination because the patient was unable to hold air well, however, P³² readings were taken and were definitely positive. Cell blocks of the ascitic fluid showed malignant cells, class V. Radioactive phosphorus was instilled into the abdomen to control the ascites, and the patient was discharged home without further treatment. He expired two months later (Fig. 48).

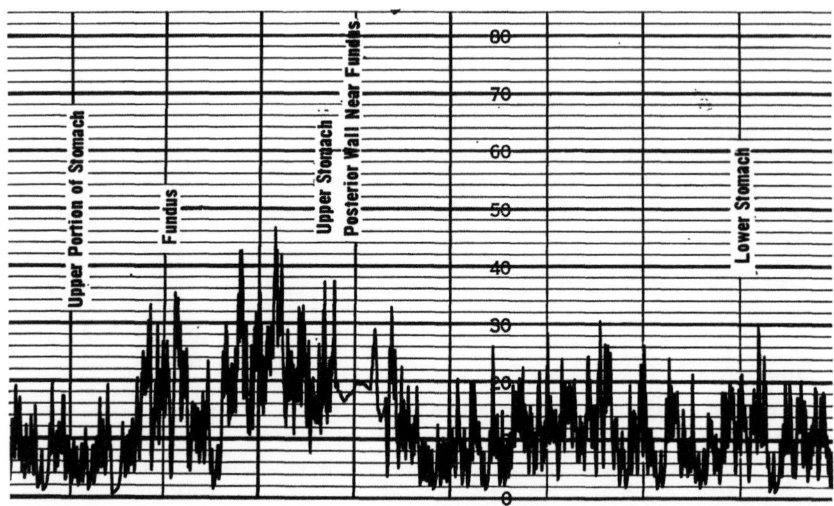

Fig. 47 (Case 26). Positive findings were recorded in the upper stomach of a patient with lymphoma. The gross appearance was not diagnostic of tumor

Case 28. A 48 year old white female had had a subtotal gastrectomy 14 months previously for benign peptic ulcer. Her symptoms on admission consisted mainly of postprandial pain in the abdomen. X-ray examination showed a 3 cm. rounded filling defect with fairly smooth margins on the lesser curvature side of the gastric pouch which was felt to be probably benign, but malignancy could not be ruled out. At gastroscopic examination, there was a well-functioning enterostomy, and no ulceration noted. However, what appeared to be submucosal tumor nodules were seen involving the lesser curvature and posterior walls of the stomach. The P³² test was positive. At laparotomy, a large section of redundant gastric mucosa was seen on the lesser curvature of the gastric pouch. Adhesions were excised as well as the redundant mucosa. Three months later, she was relatively asymptomatic. No tumor or carcinoma was found (Fig. 49).

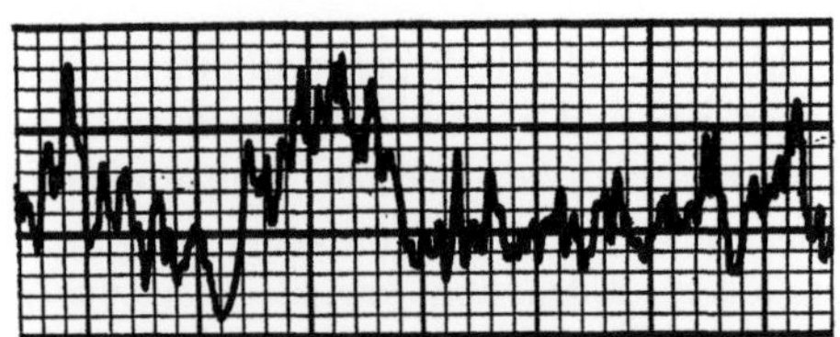

Fig. 48 (Case 27). Scattered increases in radiation were found in a patient with diffuse carcinomatous infiltration of the stomach wall. A definite diagnostic differential can be recorded, but localization of areas is poor

Case 29. A 53 year old white female had noted epigastric fullness and pain over a period of three months. There had been 30 pounds of weight loss in the past six months. X-ray examination showed large polypoid tumors, primarily extramucosal, involving the stomach. It was felt that this was more likely to be lymphoma rather than carcinoma. A gastric biopsy at the time of gastroscopy failed to show any tumor although, grossly, there was marked involvement of antrum, posterior wall and lesser curvature, by tumor. A P³² test was positive. Laparotomy was planned, but the patient became jaundiced. A liver biopsy showed atypical lymphocytic and reticulum cell proliferation in hyalinized connective tissue, suggestive of Hodgkin's disease. The patient was started on Cytoxan, but developed marked bone marrow depression necessitating discontinuance. She continued to deteriorate and expired two months after admission (Fig. 50).

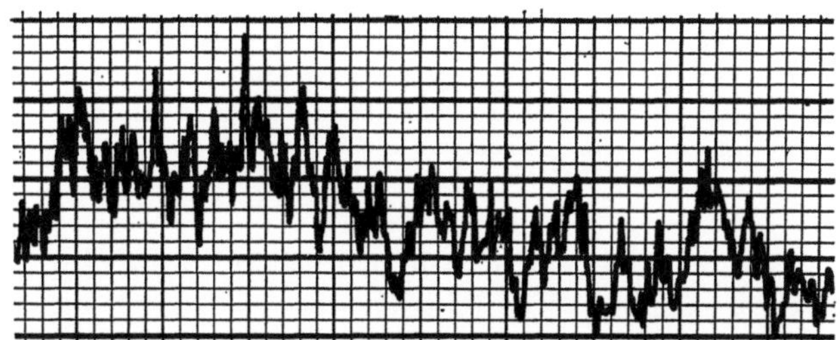

Fig. 49 (Case 28). This false positive result was recorded in a patient with postgastrectomy symptoms following resection for peptic ulcer. No tumor was found at laparotomy

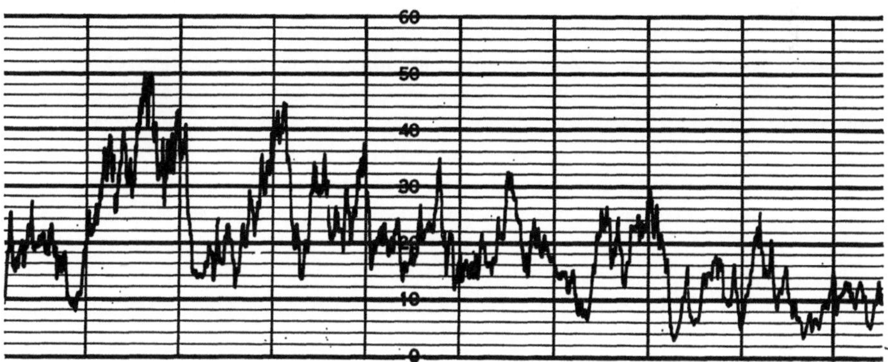

Fig. 50 (Case 29). A diagnostic differential was obtained in a patient with Hodgkin's disease of the stomach

Case 30. The patient, a 58 year old white male, had had indigestion for many years, and had taken hydrochloric acid intermittently over this period. One year prior to being seen, he began to have loss of appetite with slow weight loss, recurrent nausea and malaise. For the past three months, he had experienced epigastric pain. X-ray examination showed an extensive carcinoma of the major portion of the stomach, extending from the proximal body to the antrum, and possibly including the duodenal bulb. A Schilling test revealed 1.3 per cent excretion although the blood count was well within normal limits. At gastroscopic examination, a large round tumor mass could be seen extending from the greater curvature and anterior wall of the stomach, partly occluding the lumen of the lower stomach. P^{32} readings were within background limits. At laparotomy, massively invasive carcinoma of the stomach was noted and total gastric resection was carried out. He did well until ten months later when he had fullness in the lower abdomen and loss of appetite. A barium enema revealed a lesion in the distal part of the ascending colon consistent with a malignant process. He was given a course of 5-Fluorouracil but there was no marked response. Three months later he died of intestinal obstruction and abdominal carcinomatosis, metastatic from the stomach (Fig. 51).

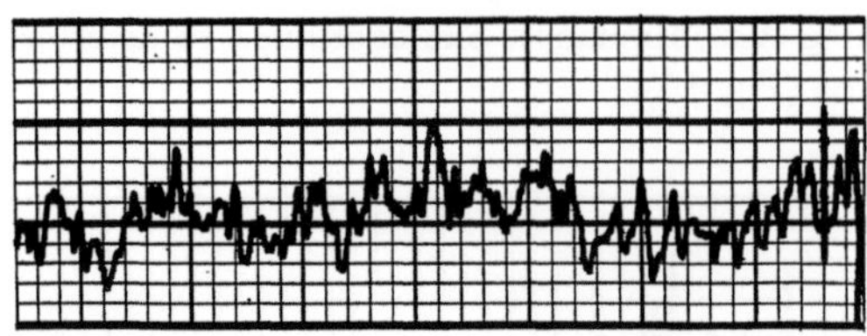

Fig. 51 (Case 30). Background radiation only, constituting a false negative reading, was noted in a patient with extensive adenocarcinoma of the stomach

Case 31. This 54 year old colored male had had epigastric distress, burning and indigestion for three months, which initially was relieved by antacids, but later became persistent. There had been a 27 pound weight loss, weakness, malaise and anorexia. Roentgenograms of the stomach showed a fungating defect of the lesser curvature in the upper portion of the body, consistent with a carcinoma. A P³² balloon study appeared to be positive. The liver was enlarged, and there was retention of bromsulphalein and elevation of alkaline phosphatase. On gastroscopic examination, there was noted to be a fixed fold extending from the junction of body and fundus on the anterior wall around close to the greater curvature. This area was nodular but submucosal, and no ulceration was present. P³² readings were all background. At laparotomy, a large carcinoma of the body of the pancreas involving the stomach, lymph nodes around the celiac axis with massive liver metastases were found. No resection was attempted. The patient died two weeks later (Fig. 52).

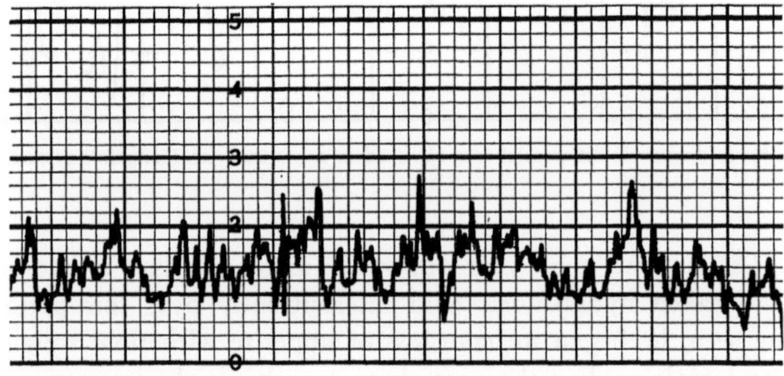

Fig. 52 (Case 31). Background radiation values were obtained in a patient with carcinomas of the pancreas which involved the stomach secondarily. The mucosa was not invaded

Case 32. This 71 year old white male had noted abdominal pain for several years, accompanied occasionally by nausea but without vomiting. At x-ray examination, a gastric ulcer was noted in the posterior wall of the body of the stomach which appeared to be benign. On gastroscopic examination, a large ulcerated lesion on the posterior wall of the stomach about 2 cm. below the fundus was seen, which appeared to be tumor infiltrating with ulceration, and was diagnosed as an ulcerating gastric carcinoma. All P³² readings were background. A P³² balloon study of the stomach was negative. At laparotomy, 60 per cent gastric resection was carried out, and the findings were benign gastric ulcer. Ulceration extended through to the serosal layer, and there was marked inflammatory change around the ulcer crater. The patient was essentially asymptomatic one and one-half years later (Fig. 53).

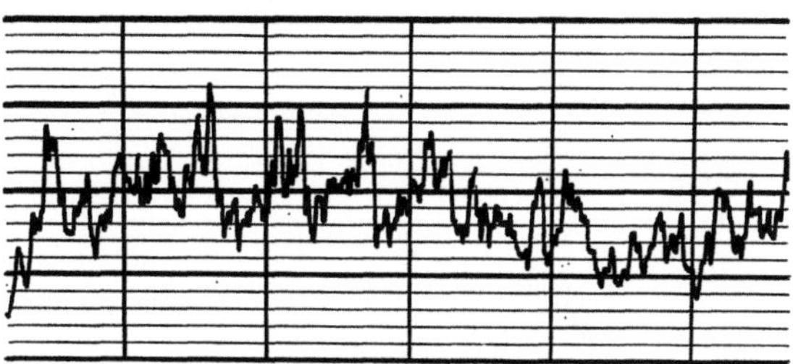

Fig. 53 (Case 32). Nondiagnostic variations in radiation from the gastric mucosa over the involved area are shown in a patient with a large benign ulcer

Case 33. A 59 year old white male who had been diagnosed, and treated as pernicious anemia for a period of several years was seen with the symptom of mild epigastric crampy pain, nausea and vomiting, but without hematemesis. X-ray examination, and gastroscopy previously, had shown a number of gastric polyps, and one which appeared to be a definite carcinoma. At gastroscopic examination, marked gastric atrophy was noted, as well as several benign looking polyps, and a 2 to 3 cm. polypoid lesion on the posterior wall which appeared to be carcinoma. P^{32} readings

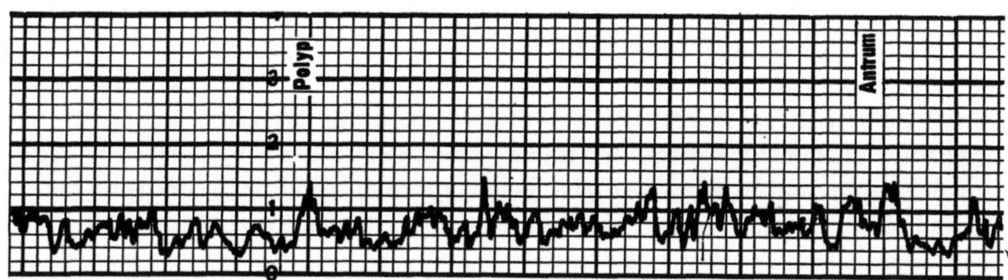

Fig. 54 (Case 33). The polyp tested in this patient with pernicious anemia, which was grossly carcinoma on gastroscopic visualization, gave only background radiation. The lesion was benign histologically when resected

were all negative over this lesion. A P^{32} test made with the sensitized balloon was also negative. A repeat roentgenogram showed an abnormality in the mucosal pattern in the antral portion of the stomach, compatible with adenocarcinoma. At laparotomy, only multiple benign gastric polyps were found; one was large and hemorrhagic, obviously the one thought grossly to be carcinoma. The patient was asymptomatic and eating well three and one-half years later (Fig. 54).

d) Rectum

Method. In testing the rectum, it is desirable to have all surfaces as clean as possible. This involves administration of castor oil the previous day, as well as enemas until clear on the morning of the test. Unless a painful lesion is present, premedication with meperidine and barbiturates is unnecessary, although in the individual case they may be useful since the P^{32} test prolongs the procedure somewhat.

Both large and small Geiger counter tubes, inserted into Teflon probes as previously described, are employed (Fig. 55). In early experience with rectal testing, only the small (25×5 mm.) tube was used. This instrument gave quite satisfactory results, and is especially valuable for scanning small lesions under direct vision through the sigmoidoscope. During the past two years, the larger probe has been used experi-

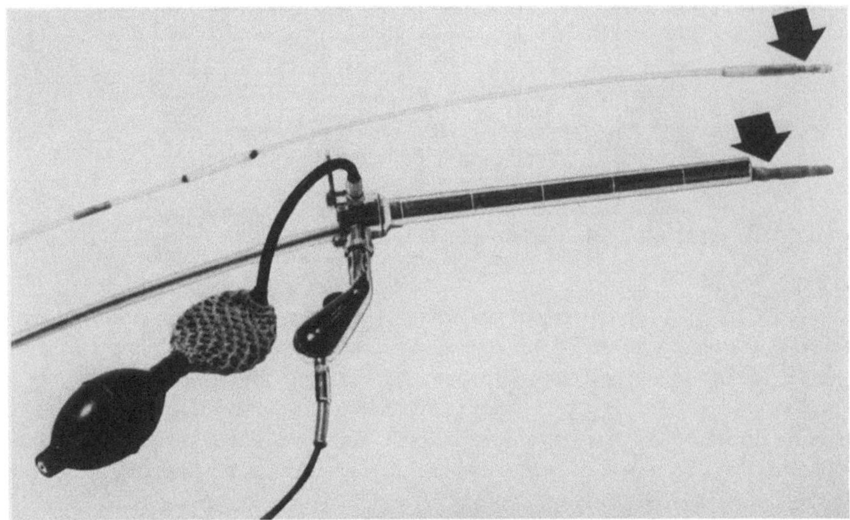

Fig. 55. Both probes used in the rectum are illustrated, the larger in position in the standard proctosigmoidoscope with the Geiger counter protruding. Distances are measured from the anal orifice in centimeters as the instrument is withdrawn and counts are taken at approximately two centimeter intervals

mentally, especially for survey of the rectum in which no gross lesions, or only apparently inflammatory areas, are found. Radiation values are appreciably greater with this probe, and it appears to give better approximation to the walls of the rectum and lower sigmoid than the small probe; hence, its value for routine scans. When the same malignant lesion is tested with each probe in turn, a diagnostic reading may be obtained with both, the differential being approximately the same. Attempts to pass the large probe beyond the 25 cm. distance possible with the average sigmoidoscope have been disappointing, and have been limited for fear of perforating the bowel. Both types of probes, employed as indicated, have been extremely satisfactory.

e) Results

Cancer. There were two false negative tests in 25 patients, for an accuracy of 92 per cent. Two of the malignant lesions were lymphomas, and 23 adenocarcinomas. Three of the adenocarcinomas developed in villous adenomas, and one in a case of familial polyposis. One of the false negative results was in a patient with lymphoma, with extremely extensive involvement (multiple biopsies in numerous areas were all positive). All readings on the P³² test with the large probe were markedly elevated, and it is possible that there was simply no normal tissue for background comparison. Unfortunately, this possibility was considered too late for comparison with extra-rectal tissue.

Benign tumors. There were 10 benign tumors tested, two of which were familial polyposis, two villous adenomas, and six adenomatous polyps. Of this group, two of the adenomatous polyps gave false positive readings, for an accuracy of 80 per cent.

Inflammatory lesions. In a group of 5 nonspecific inflammatory lesions, two patients had ulcerative colitis, one a benign ulcer, and two chronic inflammation at the site of anastomosis of a previous resection for cancer. One of the latter was tested twice. All tests in this group were correctly proven negative.

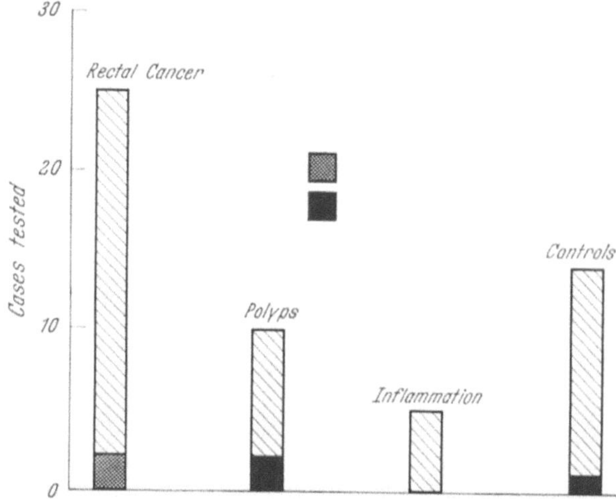

Fig. 56. The graph illustrates the excellent results obtained in rectal scanning in a relatively small number of patients. Accuracy in the rectum approaches that in the esophagus. Results of 54 rectal tests. Cancer accuracy 92%; overall accuracy 92.4%. O False negative, ● False positive

Controls. In 14 patients without pathology, there was one false positive, for an accuracy of 95.7 per cent.

Overall accuracy in 54 rectal tests involving 53 patients was 92.4 per cent (Fig. 56).

Case Reports

Case 34. A 44 year old colored female had a six months' history of purulent foul smelling bowel movements with bright red blood. Proctosigmoidoscopic examination was entirely negative, and P^{32} readings with a large probe were background in nature. Roentgenograms of the gastrointestinal tract were all within normal limits, and no cause for her symptoms was found. She was discharged to the referring physician in good condition (Fig. 57).

Case 35. This 55 year old white male complained of suprapubic pain for two years, and pain in the rectum for one year, during which time there had been some change in the size and caliber of the stool, which had become smaller. At sigmoidoscopic examination there was marked narrowing of the lumen of the rectum at a depth of 10 cm., and what appeared to be massive infiltration limiting the lumen of the organ. Three biopsies showed rectal mucosa but no tumor. Five days later, the sigmoidoscopic examination was repeated. The same gross findings were noted. P^{32} readings were taken which appeared to be positive. Biopsies on this occasion showed adenocarcinoma. Abdominal perineal resection was carried out uneventfully, revealing adenocarcinoma of the rectum with infiltrated muscular wall and tumor present in the serosa. He was doing well on follow-up three months later (Fig. 58).

Case 36. This 57 year old white female was admitted with a history of intestinal obstruction and laparotomy two months previously. At the time of previous surgery no biopsy was taken, and apparently no definite tumor had been seen, but a

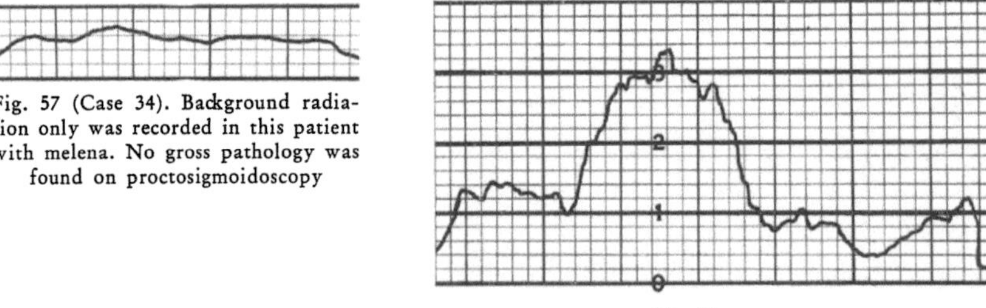

Fig. 57 (Case 34). Background radiation only was recorded in this patient with melena. No gross pathology was found on proctosigmoidoscopy

Fig. 58

Fig. 58 (Case 35). A definite radiation differential was recorded over the involved area in a patient with carcinoma, limited to the area shown. Biopsy on initial examination was negative

diverting colostomy performed. There was no true anemia. Stools were negative for blood, ova and parasites, and there was no evidence of any metastatic extension of tumor either on physical or x-ray examination. Prior to her laparotomy, her only

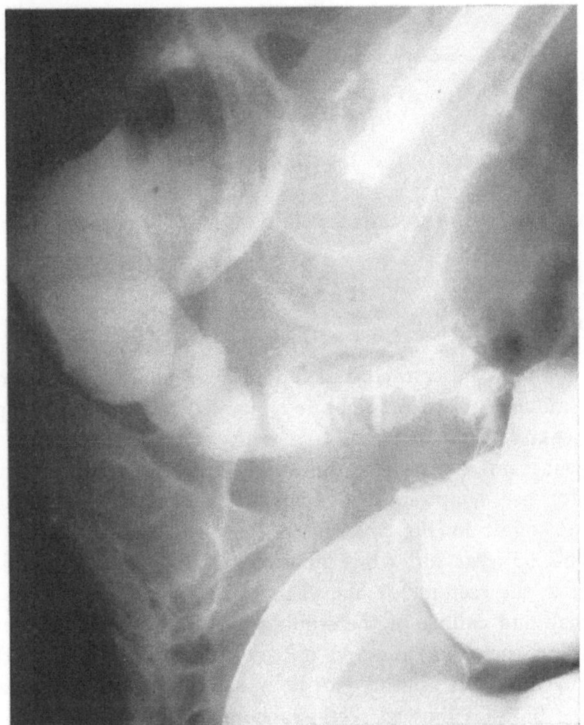

Fig. 59

symptom had been constipation. On admission, roentgenoscopy of the colon by way of colostomy and rectum showed no definite pathology of the large or small bowel (Fig. 59). On sigmoidoscopy, the instrument could be advanced only to 20 cm. because of an area of marked constriction of the lumen. There was no evidence of intrinsic tumor in the rectum. P^{32} test was markedly positive in the region of the constriction (Fig. 60). Laparotomy was again performed, and at the time of surgery

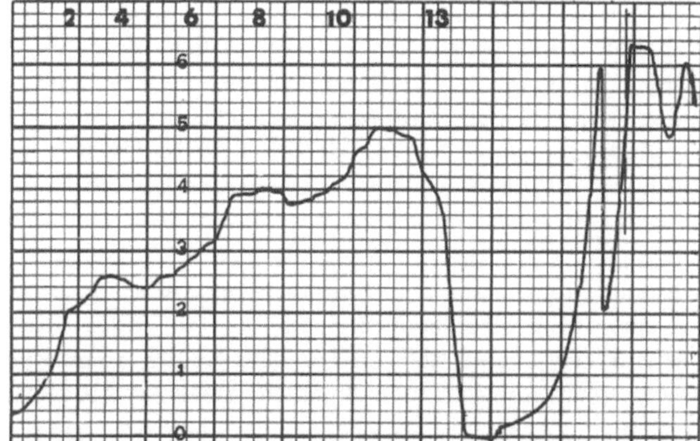

Fig. 60

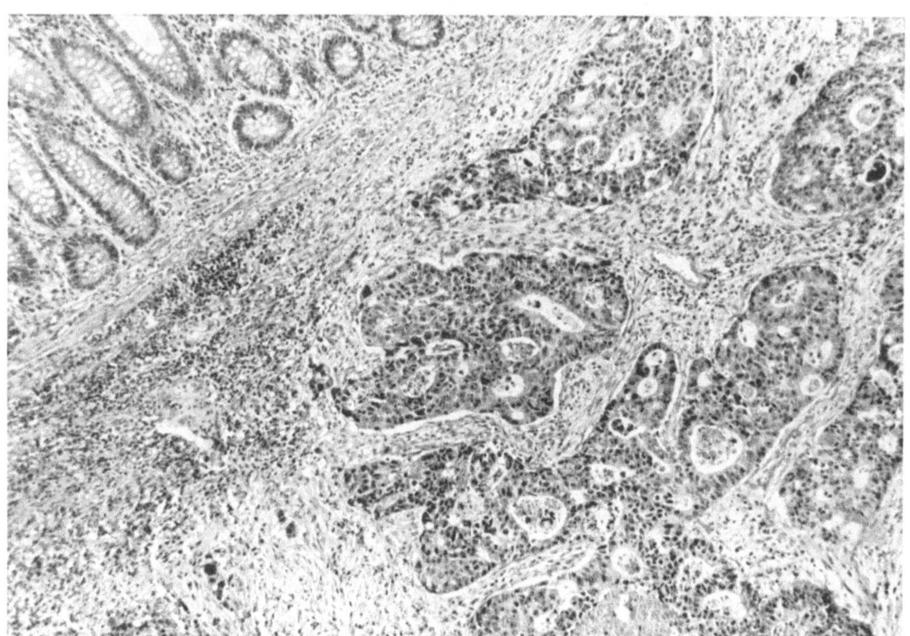

Fig. 61

Figs. 59, 60 and 61 (Case 36). Roentgenograms failed to show carcinoma of sigmoid (Fig. 59) but P^{32} test was positive (Fig. 60) and carcinoma was found at laparotomy (Fig. 61)

a definite diagnosis had not been made. It was decided to revise the colostomy, and if necessary, to do an abdominoperineal resection if tumor were found. Laparotomy revealed an adenocarcinoma of the sigmoid colon (Fig. 61). No tumor extension was noted into the lymph nodes, and an abdominoperineal resection was carried out. One month later she was doing well.

Discussion

The results obtained with the radioactive phosphorus (P³²) test in the esophagus, stomach, and rectum closely resemble those observed in other areas of the body, and are, as might be expected, subject to the same type of limitations. These consist of the relatively shallow tissue penetration of Beta particles from this source, with consequent necessity to obtain accurate approximation of the Geiger counter when used as a scanning device, as well as confusing background radiation in some inflammatory lesions, which may persist for considerable periods. The first limitation, that of occasional inability to properly place the counter, results in false negative readings, while the increased uptake in inflammatory lesions will result in false positive results.

The gastrointestinal organs tested also possess some characteristics which to some extent set them apart from other body regions. They are rather inaccessible, long, hollow tubes, with a very considerable surface. This feature is not a problem if the P³² test is reserved solely for evaluation of lesions noted by x-ray examination, in which case a small area only may be scanned, i.e., the suspicious region. However, the test in our experience has been most valuable in determining the extent of lesions previously noted, and in some instances diagnosing undetectable neoplasia, as well as differentiating benign from malignant. While in most cases it provides only supplementary evidence, in a few instances, values obtained from the P³² scan have been the deciding factor in diagnosis in patients where other objective evidence obtained by roentgenoscopy, endoscopy, or cytology was confusing or absent. Most cases of malignant neoplasia of the gastrointestinal tract are first seen in rather advanced states. Certain experiences have pointed to the possibility that the use of the test early in the course of symptoms would help to resolve equivocal findings and present the patient for surgery in a resectable state. This type of use would necessitate routine scanning in many patients where the evidence for cancer might, to say the least, be considered as equivocal. Nevertheless, the results appear to support such procedure in certain cases at the present time, and may well indicate even more routine use in the future with improved equipment and methods.

Esophagus. Large exophytic malignant tumors of the esophagus are almost uniformly positive on the P³² test. Positioning of the Geiger tube is simple unless there is narrowing above the lesion, and in most instances a little maneuvering will allow the tube to pass close to neoplastic tissue. In some cases, it is possible to pass the tip of the esophagoscope or the probe beyond the lesion, and a rather accurate delineation of its extent may be obtained (case 17). In some tumors, there may be a considerable submucosal extension beyond the base of the visible lesion, over which radiation will be diagnostic. Positive biopsies are obtained in a large per cent of such tumors, which renders the P³² test academic unless nonoperative therapy, such as radiation or chemotherapy, is contemplated. Repeat testing at intervals has been shown to be quite accurate in detecting disappearance of neoplastic tissue from the esophageal mucosa, as well as its return when treatment is no longer effective (cases 13, 14, 15). We have had no experience with the use of preoperative radiation, but the P³² test might well be employed to determine the optimum time for surgical intervention. Although the degree of accuracy with this type of tumor is given in this study as only 95.7 per cent, due to two false negative results obtained early in the study, there have been no errors for the past four years, and correct results should, in time, approach 100 per cent.

In the case of the gastric carcinoma invading the esophagus from below, testing has been unusually valuable in determining the amount of submucosal invasion as well as in making the diagnosis of neoplasia. When tumor was identified in the lower esophagus, the test was uniformly positive, although one biopsy was negative (10 patients). With no recognizable tumor, submucosal extension was diagnosed in 11 of 13 patients, although seven of ten biopsies were negative histologically. The test should be performed routinely for possible lesions of the cardioesophageal junction, and excellent accuracy may be expected.

Results obtained in patients with esophagitis have likewise been quite good. Two patients who gave false positive readings had massive accumulations of inflammatory tissue around the involved area, which may have been a factor in producing the unusually heavy radiation noted (case 22). A longer interval between P^{32} administration and scanning might reduce the incidence of false positive readings in esophagitis. Perhaps a time interval of 24 to 48 hours, instead of 18 to 20 hours would be more realistic when esophagitis is clinically suspected, even though this might be slightly more inconvenient. The majority of patients (90 per cent) will, however, give correct results with the present methods. Clinically, it is notable that a medical regimen was sufficient to control symptoms in all 20 esophagitis patients with a negative P^{32} test, whereas the two yielding false positives were resistant to the standard methods of control, and surgical intervention was mandatory for relief. Achalasia or cardiospasm (eight patients), and controls (46 patients) each yielded one false positive test, probably related to some failure in methodology.

Stomach. The relatively poor results obtained with the P^{32} test in the stomach, based on the difficulty of approximating the Geiger counter under direct vision, render it much less useful and largely experimental at the present time. The problem of accurate counter placement is the same as that of adequate mucosal biopsy, which at present is usually obtained by blind suction methods. It must be recognized that any test performed with radioactive phosphorus as the trace element in the stomach is subject to considerable chance of error unless direct vision scanning, under gastroscope control, is possible. As noted by STEIN et al. (1965), a false notion of the worth of the procedure may be gained otherwise in the negative stomachs, which will comprise a large proportion of all tests. It should also be pointed out that cases reported in several studies include patients with very large lesions. Not only are these relatively easy to scan, but such diagnoses are easily made by conventional means such as roentgenoscopy and gastroscopy, and probably mean little in the overall evaluation of the P^{32} test except to show that it is indeed possible to obtain a diagnostic differential from malignant neoplasia in the stomach. Our findings support this general conclusion, but leave open to question the much larger problem as to how valuable the test is in its present state of development and methodology. Our results would indicate that it leaves much to be desired.

Attempts on our part to duplicate the results of testing with P^{32} employing the coated balloon method of ACKERMAN et al. were even more discouraging than those of STEIN. It appeared impossible to determine exactly where the balloon was placed in the stomach, since fluoroscopy could not be used, and confusing artefacts would appear unless extreme caution was used in handling the exposed balloons. The balloons were quite expensive, and the emulsion seemed to deteriorate rather rapidly. The study was eventually terminated as being too unpromising to warrant further trials.

Although we have had no experience with the solid state detector of OTTO, this method would appear to have the same limitations as other attempts to scan the stomach after P^{32} administration. The main determining factor in this case would again appear to be the accuracy of positioning of the detector. Experience with numerous patients leads to doubt that either a Geiger counter or solid state detector can be accurately positioned on small lesions under fluoroscopic control, although this may be entirely possible with large tumors of the stomach. Once again, it must be emphasized that while the differentiation of large lesions is important to establish the validity of the test, the proof of value must rest with accurate diagnosis of small and equivocal pathological areas. Evidence presented for the solid state detector method appears no more convincing than that for the Geiger counter trials, and the claim that the test can be completed in 10 to 15 minutes is highly suspect because of the necessity to wait for radiation build up in each test location. Such short intervals might be possible if only a few sites were evaluated.

The low degree of accuracy obtained in our gastric tests represents as objective an evaluation of the procedure as it was possible to make. The rather large proportion of false positive readings with benign ulcer corresponds with those obtained with the coated balloon of ACKERMAN et al. Increased accuracy obtained in benign disease otherwise must be suspect unless the counter was adequately applied to all, or the suspected, areas of the stomach. Studies are not really comparable unless the size of the lesion is noted, as well as its availability for counting. In general, the results of most P^{32} studies in the stomach lead to the conclusion that as yet it must be considered experimental in this organ, and that little reliance can be placed on results clinically unless the lesion tested is scanned under direct vision through a gastroscope. The instruments for reliable, routine testing in this manner have not yet been devised. Our experience leads us to believe that a separately controllable probe, with Geiger counter, which may be kept under direct observation through a gastroscope and applied to any given area, will supply the answer. Such an instrument should be available in the not-too-distant future.

Rectum. The rectal mucosa is the most readily available of any of the three organs tested. Sigmoidoscopy is a relatively simple procedure, and visualization of rectum and lower colon usually excellent. There is in addition ample room for application of probe and counter, the latter in larger sizes for higher counts and extremely adequate recording. The results in testing cancerous lesions approached those in the esophagus (92 per cent). Inflammatory lesions and normal controls also gave highly accurate readings (100 and 95.7 per cent, respectively). Two adenomatous polpys out of ten, however, gave false positive results, presumptively on the basis of increased mitotic activity in these particular tumors. Villous adenomas, with and without cancerous degeneration, were accurately scanned, but the number was small.

Although in general the P^{32} test will not be necessary to detect malignancy of rectum and lower sigmoid, certain extensive benign lesions, such as nonspecific inflammation or villous adenomas may well bear scanning with considerable benefit to the patient. The numbers so far tested are small, and results must be confirmed by more extensive clinical trials.

Summary

The radioactive phosphorus (P³²) test employing the miniature Geiger counter probe has proven clinical value in the diagnosis of malignant neoplasia in the esophagus. It should be routinely employed in all questionable lesions initially, and is of value in follow-up of treated carcinoma.

Evaluation of the stomach by this means is considerably less reliable because of technical difficulties involved in precisely approximating the Geiger counter to the test area. Better equipment, not now available, may in time solve this problem. Clinical use is not at present practicable, and the method remains experimental in this organ.

The diagnosis of cancer of the rectum is for the most part so readily made during routine sigmoidoscopy as to render P³² testing academic in this area. The results of such testing are extremely accurate, and may be of use in inflammatory lesions, as well as certain benign tumors. Clinical evaluation in these areas has so far been limited, and final assessment of usefulness must await more extensive trials.

References

ACKERMAN, N. B., D. B. SHAHON, A. S. McFEE, and O. H. WANGENSTEEN: Recognition of gastric cancer by in vivo radioautography. Ann. Surg. 152, 602 (1960).

—, A. S. McFEE, and O. H. WANGENSTEEN: Refinements in technique of in vivo radioautography of the stomach. Surgery 51, 295 (1962).

— —, J. A. BLUM, E. L. MAKOWSKI, and O. H. WANGENSTEEN: Multiple-organ cancer screening with diagnostic radioautographic procedures. J. Amer. med. Ass. 183, 36 (1963).

NAKAYAMA, K.: Diagnostic significance of radioactive isotopes in early cancer of alimentary tract, especially esophagus and cardia. Surgery 39, 736 (1956).

NELSON, R. S., W. C. DEWEY, and R. G. ROSE: The use of radioactive phosphorus (P³²) and a miniature Geiger tube to detect malignant neoplasia of the gastrointestinal tract. Gastroenterology 46, 8 (1964).

OTTO, D. L., N. H. HORWITZ, R. S. KURTZMAN, and J. E. LOFSTROM: Radioactive I¹³¹ and P³² as aids in the diagnosis of lesions of the stomach. Amer. J. Roentgen. 91, 66 (1964).

SCHULMAN, J., JR., M. FALKENHEIM, and S. J. GRAY: Phosphorus turnover of carcinoma of human stomach as measured with radioactive phosphorus. J. clin. Invest. 28, 66 (1949).

STEIN, G. N., V. O. TACHDJIAN, N. MAGID, and F. VILARDELL: Evaluation of balloon radioautography in the differentiation of benign and malignant gastric lesions. Arch. int. Med. 115, 326 (1965).

VOZNYUK, E. I.: The use of radiophosphorus P³² in diagnosing stomach cancer. Un. int. Cancr. 19, 1184 (1963).

Monographs already published

Schindler, R., Lausanne: Die tierische Zelle in Zellkultur (Volume 1).

Neuroblastomas — Biochemical Studies. Edited by C. Bohuon, Villejuif (Volume 2, Symposium).

Hueper, W. C., Bethesda: Occupational and Environmental Cancers of the Respiratory System (Volume 3).

Goldman, L., Cincinnati: Laser Cancer Research (Volume 4).

Metcalf, D., Melbourne: The Thymus. Its Role in Immune Responses, Leukaemia Development and Carcinogenesis (Volume 5).

Malignant Transformation by Viruses. Edited by W. H. Kirsten, Chicago (Volume 6, Symposium).

Moertel, Ch. G., Rochester: Multiple Primary Malignant Neoplasms. Their Incidence and Significance (Volume 7).

New Trends in the Treatment of Cancer. Edited by L. Manuila, S. Moles and P. Rentchnick, Genève (Volume 8).

Lindenmann, J., Zürich / P. A. Klein, Gainesville, Florida: Immunological Aspects of Viral Oncolysis (Volume 9).

Nelson, R. S., Houston: Radioactive Phosphorus in the Diagnosis of Gastrointestinal Cancer (Volume 10).

In production

Freemann, R. G., and J. M. Knox, Houston: Treatment of Skin Cancer (Volume 11).

Lynch, H. T., Houston: Hereditary Factors in Carcinoma (Vol. 12).

In preparation

Anglesio, E., Torino: The Treatment of Hodgkin's Disease.

Chiappa, S., Milano: Endolymphatic Radiotherapy in Malignant Lymphomas.

Denoix, P., Villejuif: Le traitement des cancers du sein.

Fisher, E. R., Pittsburgh: Ultrastructure of Human Normal and Neoplastic Prostate.

Fuchs, W. A., Bern: Lymphography and Tumordiagnosis.

Grundmann, E., Wuppertal-Elberfeld: Morphologie und Cytochemie der Carcinogenese.

Hayward, J. L., London: Cancer of the Breast: Hormonal Changes.

Irlin, I. S., Moskva: Mechanisms of Viral Carcinogenesis.

Kern, G., Köln: Carcinoma in situ.

Koldovsky, P., Praha: Transplantation Tumor Specific Antigen (TTSA).

Langley, F. A., Manchester: Epithelial Abnormalities of the Cervix Uteri.

Martz, G., Zürich: Hormonbehandlung der Tumoren.

Mathé, G., Villejuif: L'Immunotherapie des cancer.

MEEK, E. S., Bristol: Antiviral and Antitumour Agents of Biological Origin.

NEWMAN, M. K., Detroit: Neuropathies and Myopathies Associated with Occult Malignancies.

ODARTCHENKO, N., Lausanne: Prolifération cellulaire érythropiétique.

PACK, G. T., New York: Clinical Aspects of Cancer Immunity and Cancer Susceptibility.

PACK, G. T., New York / A. H. ISLAMI, New York: Tumors of the Liver.

RITZMAN, S. E., Galveston / W. C. LEVIN, Galveston: The Syndrome of Macroglobulinemia.

STEWARD, J. K., Manchester: Tumors in Children.

WEIL, R., Lausanne: Biological and Structural Properties of Polyoma Virus and its DNA.

WILLIAMS, D. C., Caterham, Surrey: The Basis for Therapy of Hormon Sensitive Tumours.

WILLIAMS, D. C., Caterham, Surrey: The Biochemistry of Metastasis.

ZILBER, L. A., Moskva: Virogenetic Theory of Cancer Origin.

Herstellung: Konrad Triltsch, Graphischer Betrieb, Würzburg